This book belongs to:

Mr. Vincent D. Mahoney

SECRETS OF
INFINITY
150 ANSWERS TO *an* ENIGMA

Antonio Lamúa, editor

FIREFLY BOOKS

A FIREFLY BOOK

Published by Firefly Books Ltd. 2013

First printing

Publisher Cataloging-in-Publication Data (U.S.)

A CIP record for this title is available from the Library of Congress

Library and Archives Canada Cataloguing in Publication

A CIP record for this title is available from Library and Archives Canada

Published in the United States by
Firefly Books (U.S.) Inc.
P.O. Box 1338, Ellicott Station
Buffalo, New York 14205

Published in Canada by
Firefly Books Ltd.
50 Staples Avenue, Unit 1
Richmond Hill, Ontario L4B 0A7

Printed in Canada

This title was developed by:
LOFT Publications, S.L.
Via Laietana 32, 4°, of. 92
08003 Barcelona, España
Tel.: +34 93 268 80 88
Fax: +34 93 268 70 73
loft@loftpublications.com
www.loftpublications.com

Editor: Antonio Lamúa
Editorial coordination: Claudia Martínez Alonso
Assistant to editorial coordination: Ana Marques
Art direction: Mireia Casanovas Soley
Graphic edition: Manel Gutierrez (@mgutico)
Texts: Silvia Serrano, Natalia Hutter
Translation: Cillero & de Motta
Cover layout: María Eugenia Castell Carballo
Layout: Yolanda G. Román, Kseniya Palvinskaya

INTRODUCTION

The concept of infinity is a highly complex enigma, and its meaning has been considered by many disparate approaches. Thus the best method to elucidate the topic for the general reader is to explore infinity as it is studied by particular fields, including science, mathematics, technology, art, philosophy and symbology.

One useful starting point, one we find at the heart of any discussion of infinity, is the notion that what is infinite is never-ending, although beyond that starting point the arguments and ideas diverge in seemingly endless ways.

In antiquity, the idea of infinity was studied by classical philosophers from Greece and Rome, such as Zeno of Elea, Parmenides, Archimedes and Pythagoras. Aristotle was the primary figure who, around 350 BC, established a school of thought regarding infinity that would be further articulated by succeeding scholars.

One of Aristotle's greatest contributions to the study of infinity was the idea that "The whole is greater than the sum of its parts." In more recent times, mathematician Georg Cantor used set theory to further explore this concept, now known as Cantor's Theorem.

Yet no single theorem or thinker has proven capable of forming a "complete" vision of infinity. Analysis of multiple approaches from varied sources is necessary to begin to capture its true scope.

This book offers 150 different angles from which to consider the unresolved enigma that is infinity. And though the discussion remains unresolved, this exploration is sure to be illuminating.

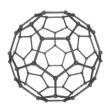

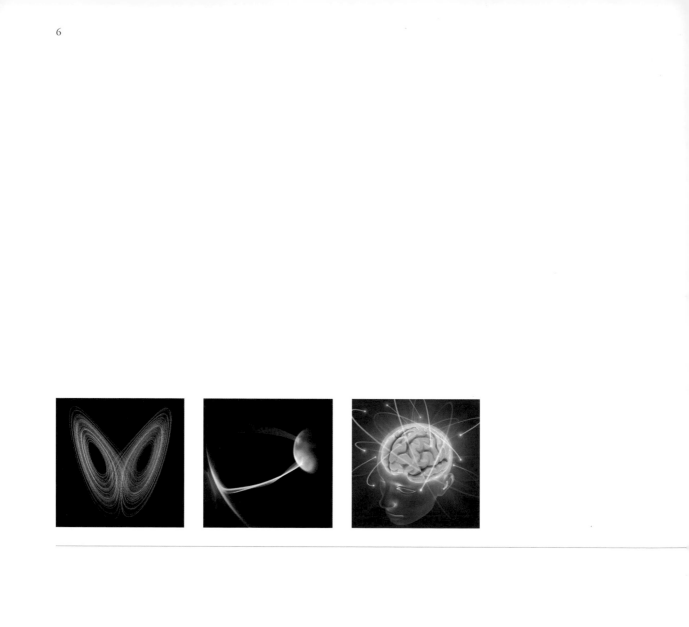

SCIENCE

SCIENCE

COMPONENTS OF LIFE ON EARTH MAY HAVE ARRIVED HERE FROM THE INFINITE

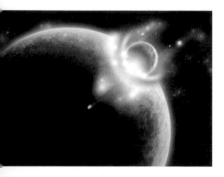

In January 2004, NASA's *Stardust* probe gathered microscopic traces of glycine collected from the tail of the comet Wild 2 in the depths of the Solar System, about 242 million miles (389 million km) from Earth. This was the first time that glycine (the simplest naturally occurring amino acid and a constituent of most proteins) was found in a comet, and the discovery supports the theory that components of life reached Earth from outer space. It is believed that comets like Wild 2—named after astronomer Paul Wild—contain well-preserved grains of material as old as the Solar System, dating back thousands of millions of years and containing clues about the formation of the Sun and the planets.

Research on the origins of life initially focused on the production of amino acids from organic materials already present on the planet. However, subsequent research showed that ancient Earth's atmospheric conditions were very poor. The atmosphere consisted mainly of carbon dioxide, nitrogen and water. Several experiments and calculations showed that the synthesis of organic molecules necessary for the production of amino acids could not occur in this environment.

Glycine is one of the twenty amino acids commonly found in proteins on Earth. These amino acids form chains that are wound together to in turn form proteins, which are the complex molecules that regulate chemical reactions inside living organisms. Scientists have long tried to figure out whether these complex organic compounds originated on Earth or in space. These latest findings lend credence to the idea that extraterrestrial objects, such as meteorites and comets, may have brought these essential ingredients of life to Earth and the other planets from elsewhere in the cosmos.

The Stardust mission is the first attempt to collect space dust beyond the moon. The estimated age of these particles dates back to the origins of the Solar System.

EINSTEIN AND PARALLEL UNIVERSES

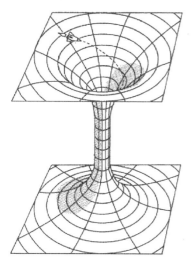

In 1915, Albert Einstein (1879–1955) presented the theory of general relativity, which reformulated the whole concept of gravity. One consequence was the emergence of the scientific study of the origin and evolution of the universe by the branch of physics called cosmology.

In modern physics there are two main theories: first, general relativity, which describes the behavior of large objects like planets, stars and matter in general, in addition to the mechanism of gravity, and secondly, the developments and theories of quantum physics, which describes the behavior of microscopic particles such as quarks, atoms, photons and neutrons, as well as the behavior of the other forces such as electromagnetic and nuclear. Yet contradictions arise when an attempt is made to bring together aspects from these two theories; their unification seems impossible, given that each one describes one part of the universe but is unable to describe the other.

Einstein's theory of relativity first predicted the existence of four-dimensional space-time and black holes. However, in 1935, Einstein and his colleague from Princeton University, Nathan Rosen, presented their new theory describing the operation of black holes. They proposed that rather than being a simple hole or crack in space-time, as initially believed, the black hole was actually a bridge connecting one universe to another universe. Einstein and Rosen argued that black holes were "bridges" to worlds and epochs unknown. This concept is known as the Einstein-Rosen Bridge.

The Einstein-Rosen Bridge was the first widely accepted scientific theory about the possible existence of parallel universes or dimensions. The work of Einstein and Rosen paved the way for subsequent generations of physicists to explore in depth the concept of parallel universes. For example, it decisively influenced the work of the "many-worlds interpretation" presented in 1957 by physicist Hugh Everett III (1930-1982). Everett's theory suggests that, in addition to our own, there exist many worlds or universes that continually divide into mutually inaccessible separate and distinct dimensions. According to Everett, each world or dimension contains a different version of the same people carrying out various actions in the same moment in time.

The hypothesis of the parallel universes has been an inspiration for numerous works of science fiction in both the world of literature and the movies.

THE NEUROPLASTICITY OF THE HUMAN BRAIN

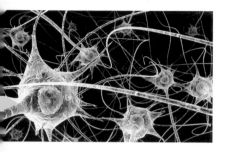

The term neuroplasticity was coined in 1948 by Polish neurophysiologist Jerzy Konorski (1903–1973), though the central idea, that the brain undergoes infinite morphological changes in response to constant external stimuli, was initially conceived by Santiago Ramón y Cajal of Spain (1852–1934).

Neuroplasticity is the ability to change the type, form, function and number of synapses (neuron-neuron connections) that link up the neural circuits as a result of experience, that is, changes in the external environment, in the body or caused by injuries. The concept is based on the process of modifying the activity and organization of neurons that produce permanent or long-term changes in the power of these synapses.

During the 20th century, neuroscientists believed that the structures of the brain were static. Our anatomy was thought to be immutable, in the sense that once we reached adulthood the only changes possible were those of steady decline. It was believed that when neuronal cells stopped functioning properly, due to being dead or damaged, they could never be replaced.

Today we know that brain connections vary throughout adult life, and that it is also possible to generate new neurons in areas related to memory management. According to the science of neuroplasticity, mental processes (such as learning and recalling) are capable of altering the pattern of brain activation in neocortical areas. Thus, the brain is not an immutable structure, but responds to the individual's life experiences, and this flexibility, these changing responses, are what cause the system to adapt.

Neuroplasticity is a key concept in recent decades that has elucidated the way the brain organizes itself in response to environmental stimuli.

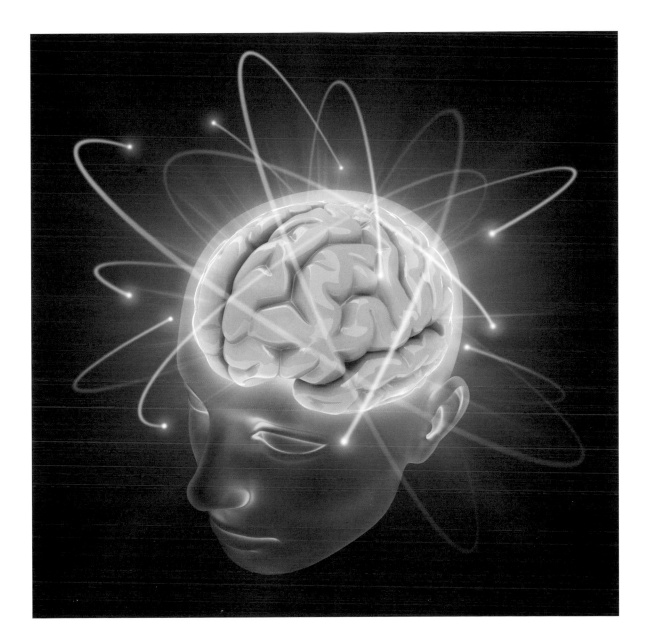

PASCAL'S TRIANGLE

In 1653, the philosopher and scientist Blaise Pascal published the *Traité du triangle arithmétique*, in which he described many of the properties of this famous triangle. This figure had already been studied by other mathematicians in Europe, but also in Eastern countries such as China, India and Persia, by the famous Al-Karaji and astronomer and poet Omar Jayyam (1048-1131) for example, five centuries before Pascal explored its applications. In China, it is called the Yang Hui triangle in honor of the mathematician who described it as early as 1303.

Pascal's triangle is composed of integers that progress infinitely in a symmetrical pattern. It starts with a 1 in the top row and numbers are placed in subsequent rows so that each is the sum of the two numbers above it. It is assumed that positions outside the triangle contain zeros, so that the edges of the triangle are formed by ones. Obviously, the triangle has no end: the rows can be extended to infinity, as the numerical values can be increased indefinitely.

Its importance lies in its varied applications in algebra, forming a small mathematical universe unto itself. It hides a variety of properties and curiosities of immense value to mathematics. For example: the sum of each row is equal to twice the sum of the previous row, the sum of each row is equal to 2 raised to the power of the order of the row (the first row corresponds to order 0), each row determines the coefficients that are obtained by developing the binomial $(a+b)^n$ (known as Newton's binomial) and each triangle number represents the value of a combinatorial number (if n is the column and m is the row, the value corresponds to the combinations of m elements taken from n to n). Prime numbers and even the Fibonacci series are also contained in the pattern.

In Italy, it received the name of the mathematician Niccolò Tartaglia, who described it in a treatise in the first half of the 16th century. In France and later in the Anglo-Saxon territories, it received the name of Pascal, who used it a century later in his studies on probability.

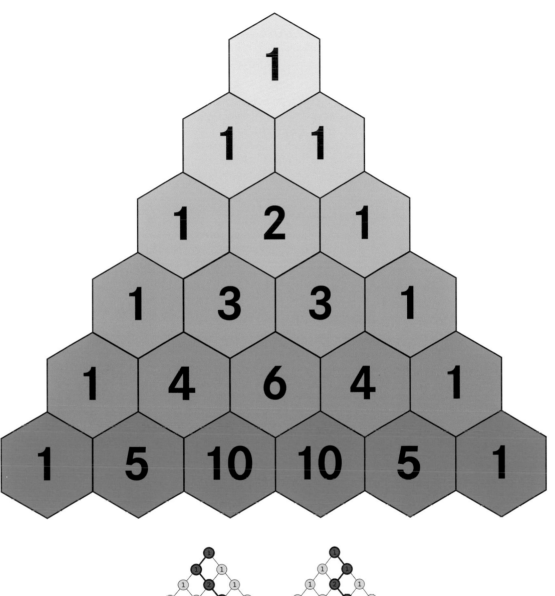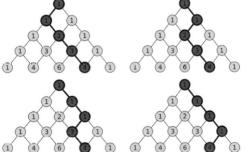

THE OSCILLATING UNIVERSE AND THE BIG BOUNCE

How the Big Crunch theory works

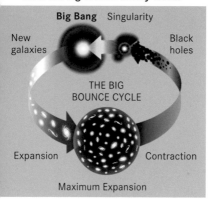

The oscillating universe hypothesis was proposed by a physical chemistry and mathematical physics professor Richard C. Tolman (1881-1948). According to this theory, the universe undergoes an infinite series of oscillations, each of them beginning with a Big Bang and ending with a Big Crunch. After the Big Bang, the universe expands for a while before the gravitational attraction of matter produces an advance until it collapses and then immediately sustains a Big Bounce.

Thus, the universe may consist of an infinite sequence of finite universes, or in other words, for every finite universe that disappears with a Big Crunch, a Big Bang would occur for the creation of the next universe. This suggests that we could be living in the first of all universes, in the two billionth universe or in any of an infinite sequence of universes.

The Big Crunch theory is a symmetrical point of view of the ultimate fate of the universe: the Big Bang resulted in an asymmetric cosmological expansion of spatial origin and the average density of the universe is enough to stop this expansion and initiate contraction. The cyclic model specifies that the Big Bang is preceded immediately by the Big Crunch of a preceding universe, creating what might be called an oscillating universe.

This hypothesis was widely accepted for a period of time by cosmologists who thought that some force should prevent the formation of gravitational singularities. However, in the 1960s, Stephen Hawking, Roger Penrose and George Ellis showed that singularities are a universal feature of the cosmos that include the Big Bang without avoiding any of the elements of general relativity. Theoretically, the oscillating universe is not consistent with the second law of thermodynamics: entropy would increase in each oscillation so as not to return to previous conditions. Other hypotheses also suggest that the universe is not limited, and demonstrate that our universe is heading for a Big Freeze or heat death, rather than a Big Crunch. However, this does not exclude the possibility that our Big Bang is preceded by the latter Big Crunch. Hawking's arguments caused cosmologists to abandon the oscillating universe model, but the theory has resurfaced in recent years in cosmology as the cyclic model.

The oscillating theory of Richard Tolman proposes that our universe is the last in a series, that is, that the universe did not have a single origin, but it has been continuously created and destroyed.

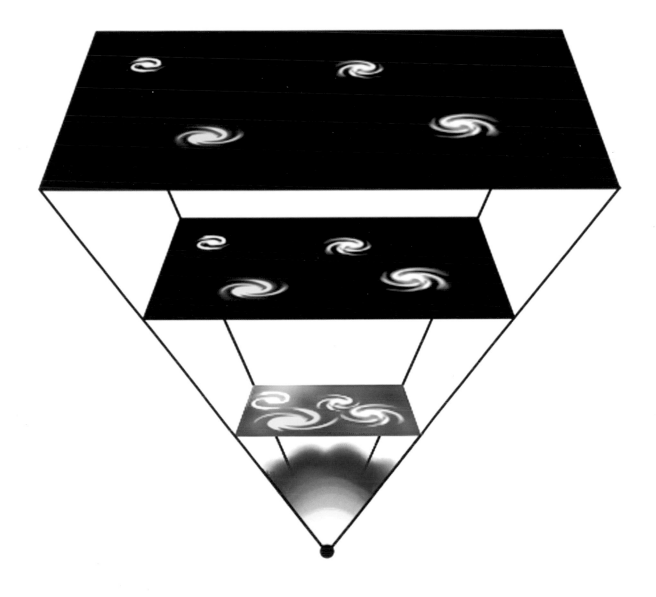

RAYLEIGH–JEANS ULTRAVIOLET CATASTROPHE

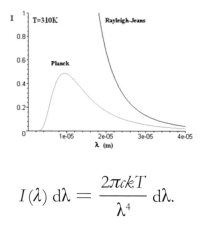

$$I(\lambda)\,\mathrm{d}\lambda = \frac{2\pi c k T}{\lambda^4}\,\mathrm{d}\lambda.$$

In the late 19th century a fundamental problem had scientists baffled: using classical physics, it was impossible to give a sensible explanation of black body radiation, an ideal solid capable of absorbing all the radiation that it receives, including visible light, so it actually looks black. However, this object emits invisible electromagnetic radiation, whose frequencies (and corresponding wavelengths) depend on its temperature and its composition.

According to the predictions of classical theory, an ideal black body should radiate in all frequency ranges, so that with increasing frequency, the energy emitted should provide an exponential increase with decreasing wavelength. In 1900, British theoreticians John Rayleigh (left) and James Jeans experimented to study the distribution of the energy of radiation emitted by a black body according to their wavelengths, and they were somewhat surprised by the result. The theory was upheld for low frequencies (infrared), however, in the ultraviolet region that continues to increase the frequency at a given point, the energy radiated ceased to increase only to decrease until the end tending to zero. In short, the energy emitted reached a maximum around 2,000 nanometers and decreased for both large and small wavelengths, which was in total disagreement with the observations of classical physics theories. This was known as "ultraviolet catastrophe," as it was in these low wavelengths (UV region) where the apparent contradiction was found between the experimental facts and what could be inferred from electromagnetic theory.

It was a new physics, called quantum, that resolved the ultraviolet catastrophe, led by Max Planck, who proposed a revolutionary idea: the discontinuity of energy. Planck proposed that energy does not emit or absorb continuously, but in packets or quanta, multiples of the frequency of radiation, and gave a mathematical formula (Planck's law) that was consistent with the empirical results obtained in the black body radiation. In this formula, he introduced his constant, h, which relates the energy of a photon with relation to the frequency of the wave, f, by the famous formula $E=h\,f$.

It is said that when Max Planck discovered his law, he was actually hesitant to accept it, and he never imagined its scope until his still relatively unknown friend, Albert Einstein, succeeded in convincing him.

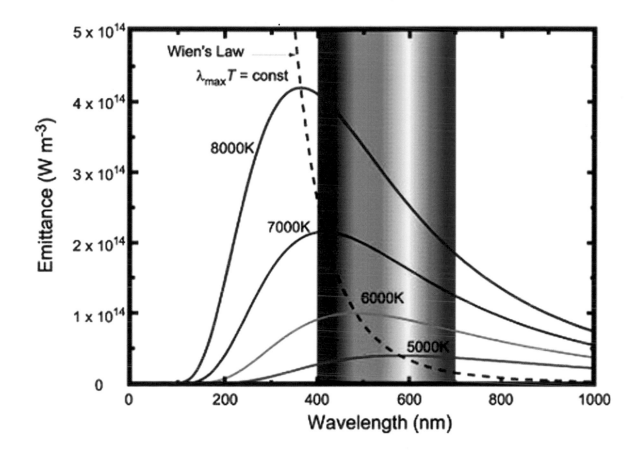

CORNU SPIRAL

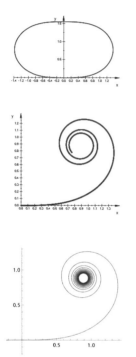

Marie-Alfred Cornu (1841–1902) was a French physicist, a student of the École Polytechnique in Paris, where he later worked as professor of experimental physics beginning in 1867.

Although he made several forays into other branches of physical science—including, with his colleague Jean Baptistin Baille in 1870, a repeat of Henry Cavendish's experiment to determine the gravitational constant G—his work primarily focused on optics and spectroscopy. In particular, he carried out a redetermination of the speed of light (after Fizeau's method), which introduced several improvements that greatly contributed to the accuracy of the results. For this achievement he was awarded the Lacaze prize, membership into the French Academy of Sciences and the Rumford Medal of the Royal Society of England.

Cornu's name is given to this famous spiral curve, though predecessors had begun to address the concept earlier, such as Euler in 1744, who used it to solve a problem proposed by Bernoulli (who had studied it around 1696). Cornu used the spiral curve in his studies on the diffraction of light.

It is a flat curve in a double spiral, with central symmetry. From its point of origin (point 0) of zero curvature and infinite radius, the radius of curvature decreases as one moves through the two arms so that the product between the radius of curvature and the distance measured along the curve remains constant. Thus, the two arms of the spiral twist and tend to converge at two predictable points of zero radius, where they will go after covering an infinite distance, after infinite turns.

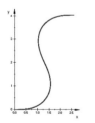

Here is the Cornu spiral's mathematical equation:

$$\rho \times s = C$$

where:
ρ is the radius of curvature, s is the development or arc, and C is the constant of the spiral.

One fascinating property of the Cornu spiral is that its curvature at any point is proportional to the distance along the curve measured from the origin. This makes it useful as a transition curve in planning highways or railway tracks, as a vehicle that follows the curve at constant speed will have a constant angular acceleration. Also, sections of the spiral clothoids are commonly used on roller coasters.

This spiral can be found in highway junctions or in links between straight sections of roads, since it can go from a straight line (infinite radius) to a circumference (finite radius) without creating centrifugal force.

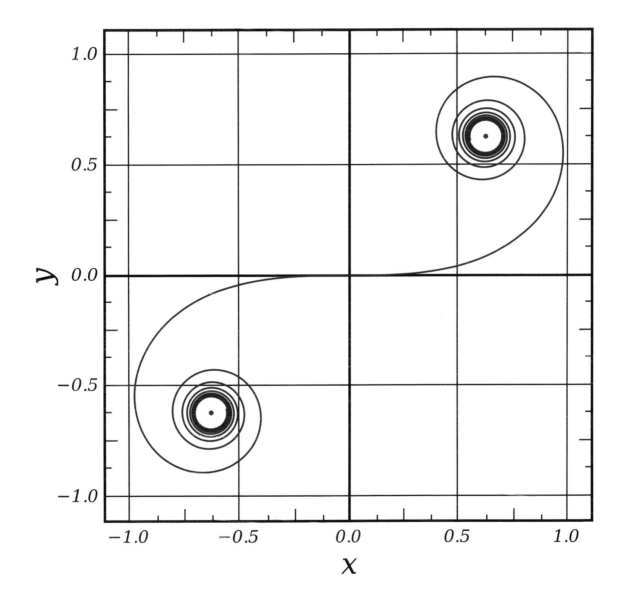

THE MOST REMOTE AND DEEPEST IMAGE OF THE UNIVERSE

The majority of studies, research and theories agree that the explosion that gave birth to the universe, the so-called Big Bang, occurred 13,700 million years ago.

The image seen at right allows for the contemplation of the galaxies as they were arranged almost at the inception of the universe. The image is of a scene millions of light years away, a depiction of the origins of creation, representing profound depths of existence never before seen. It belongs to a series of images that provides scientific data on the formation and early history of the universe.

Principal photography for this NASA mission was taken by a unique camera with a WFC3/IR wide angle lens installed during the last servicing mission of the space shuttles to the *Hubble* telescope. According to NASA, its high capacity enables it to penetrate the most remote and distant regions of the universe.

The light coming from the photographed stars travels at 186,000 miles per second (300,000 km/sec), the speed of light in space. The image is but a glimpse of the distant past, a moment that occurred billions of years ago. The elements that can be observed such as galaxies, stars or pulsars are no longer in the same place today.

The Wide Field Camera 3 (WFC3) is the most technologically advanced instrument of the Hubble *Space Telescope, and is used to take pictures of the visible spectrum. It was installed to replace the Wide Field Planetary Camera 2 in May 2009.*

OLBERS' PARADOX

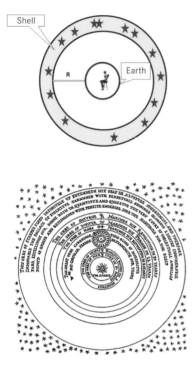

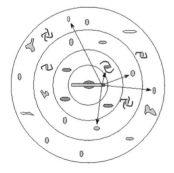

Olbers' paradox is named after the German physician and astronomer Heinrich Wilhelm Olbers, who published his treatise in 1823. The paradox resulted from the inconsistencies of Newtonian cosmology. In Newton's view, the universe is completely static, that is, flat (Euclidean) and without beginning. This stasis, however, entails that all stars in the universe would be distributed in an inifinitely uniform arrangement.

At the heart of the paradox is the contradiction between the fact that the night sky is black and that the universe is infinite. If both assertions were true, every sight line from the Earth would end in a star, so the sky should be completely bright. However, astronomers know that the space existing between stars is black.

The paradox between a dark night and an infinite universe was known before it was discussed by Olbers. In the early 17th century, the astronomer Johannes Kepler actually used the paradox to support his theory that the universe is infinite. In 1715, British thinker Edmund Halley identified some bright areas in the sky, and suggested that the sky did not light up uniformly during the night because, although the universe is infinite, the stars are not evenly distributed. In 1743, Jean-Philippe Loys de Chéseaux suggested that either the sphere of the stars was not infinite, or the light intensity decreased with distance, perhaps because of some absorbent material in space.

Almost a century later, Olbers proposed that the sky was dark because there was something in space that blocked most of the starlight that should have reached the Earth, a theory that scientists have denied, arguing that if there was matter that blocked the light, it would heat up over time and eventually radiate as brightly as the stars.

In 1948, astronomer Hermann Bondi posited that the expansion of the universe caused the light perceived from afar to be red and, therefore, it had less energy in each photon or particle of light. Finally, in the 1960s, Edward Harrison showed that the sky is dark at night because we do not see the stars that are infinitely far away. This solution depends on whether or not the universe has an infinite age, since the light takes a certain time to reach the Earth. Each sight line from the Earth would not end in a star because the light from the more distant stars necessary to create Olbers' paradox has not yet reached the Earth, that is, during the lifetime of the universe the stars have not given off enough energy to make the sky shine at night.

Heinrich Wilhelm Olbers (1758–1840) made significant scientific progress, including the discovery of the asteroids Vesta and Pallas (the third largest of the Solar System) and the development of an ingenious method to calculate the orbits of comets.

THE *HUBBLE* TELESCOPE

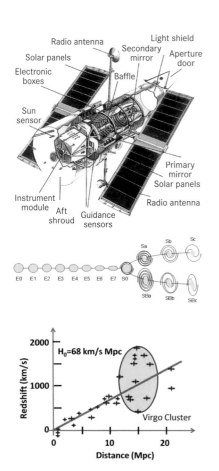

During the last 20 years the *Hubble* Space Telescope has revolutionized the way humanity looks at the infinite universe, and in many ways has become the most influential object since Galileo observed the night skies four centuries ago.

Named after the astronomer Edwin Hubble, it was launched on April 24, 1990 as a joint project of NASA and the European Space Agency. *Hubble* was conceived as a space telescope that could be visited by a space shuttle, mainly to repair damaged items, install new instruments and maintain the telescope's orbit.

An initial mistake in polishing the primary mirror of the telescope caused the initial images to be slightly blurry. Although this mistake was considered in its day a major negligence on behalf of the project, the first servicing mission to the space telescope that was carried out with the *Endeavour* shuttle in December 1993 installed an optical correction system capable of fixing the defect of the primary mirror. After the fifth and final servicing mission, carried out in 2009, it was estimated that *Hubble* would continue in service until at least 2014, when the launch of the *James Webb* telescope is planned.

From that first maintenance mission, *Hubble* has proven to be an unparalleled instrument, capable of carrying out observations that continually affect our ideas about an expanding universe. Among the most important astronomic discoveries, what scientists have called "dark energy" stands out. Solving the mystery of dark energy could revolutionize physics, triggering new theories about the origin of the universe, even resolving some speculation with regard to the final destination. *Hubble* has also allowed us to closely observe the life cycle of stars with many of its images, and perhaps one of the most famous is of the so-called "pillars of creation," columns of gas in the Eagle Nebula (M16) where the formation of new stars is taking place.

The importance of *Hubble* is unquestionable and unprecedented. Surely as we further analyze its findings, this "starry messenger" will yield further discoveries of great significance.

Edwin Hubble (1889–1953), who lent his name to the satellite, was one of America's most important astronomers of the 20th century and is considered the father of observational cosmology.

SCIENCE   **WILLIAM PEPPERELL MONTAGUE AND ETHER**

The philosopher William Pepperell Montague (1873–1953) suggested that space was static and infinite in all directions, but its astonishing dimensions were offset by its simplicity. Montague supposed that space was filled with an immobile, invisible and continuous substance, called ether (also aether, luminiferous ether), leading light waves from star to star and from atom to atom.

Before Einstein debuted his famous theory of relativity, there was an article in the *Encyclopedia Britannica* by Scottish physicist James Clerk Maxwell (1831–1879) about the existence of ether which read: "There can be no doubt that the interplanetary and interstellar spaces are occupied by a substance or material body, which is certainly the greatest of all the bodies that we know of."

It was thought then—since the speed of light depends on the density of the medium, with speed decreasing when moving through denser mediums—that ether should have a negligible density and a large coefficient of elasticity. Ether, named that for its similarities to the hypothetical substance proposed by Aristotle, was believed to be the medium through which light waves propagated in space.

It was assumed that with the appropriate experiments we could discover the direction and speed that our planet and the Solar System moved in with respect to the stationary ether. It was believed that space was composed of matter and energy; infinite time, which was considered so fundamental as to be "composed of" only time itself, was thought to be connected to space yet not a component thereof. It was believed that when two bodies collide and rebound, some part of their movement or energy passes through the ether, which is dissipated in a waveform of light or radiant heat. Yet this line of reasoning proved to be invalid.

The interpretation of the negative result of the 1887 experiment by Albert Michelson (1852-1931) and Edward Morley (1838-1923), and repeated by Georg Joos (1894-1959) in the late 1920s, to prove the existence of ether eventually dissipated the concept and provided the basis for the formulation of Einstein's theory of special relativity.

For Aristotle (384–322 BC), ether was the material element of which the supralunar world was composed, whereas the sublunary world consisted of the famous four elements: air, fire, water and earth.

THE PARADOX OF THE UNSTOPPABLE FORCE

The paradox of the unstoppable force is a self-contradiction that entertains the idea of what would happen if an unstoppable force collided with an immovable object. It is included in the book by Isaac Asimov *One Hundred Basic Questions About Science*, which includes the answers that the scientist gave readers in his section "I. A. Explains" in the journal *Science Digest*.

The common responses to this paradox appeal to either logic or semantics. Following basic logic, the immovable object and the unstoppable force are both tacitly assumed to be indestructible. It is also assumed that the two entities are separate, because an irresistible force is implicitly an immovable object, and vice versa. The paradox evidently arises based on the fact that its two premises (that there exist such things as irresistible forces and immovable objects) cannot be simultaneously true. If an irresistible force were to exist, its existence would entail that there cannot be any such thing as an immovable object, and vice versa. As for semantics, if there was such an unstoppable force then speaking of an immovable object is nonsense in this context and vice versa, it would be like asking for a four-sided triangle or what would happen if two plus two resulted in five.

The paradox is conceived as an exercise in logic, not as the postulate of a possible reality. According to modern scientific beliefs, there are not, and in fact there cannot be, unstoppable forces or immovable objects. An immovable object should have an infinite inertia and therefore, an infinite mass. Such an object would collapse under its own gravity and would create a singularity. An unstoppable force would imply an infinite energy, which, according to Albert Einstein's famous equation $E=mc^2$, would also imply an infinite mass.

Another approach to this paradox is that the object would exist perpetually, since by definition an unstoppable force is an immovable object.

Isaac Asimov (1920–1992) was a biochemist and writer who excelled in the genre of science fiction and popular science, and is widely known as the father of the laws of robotics.

THE VACUUM FLUCTUATION

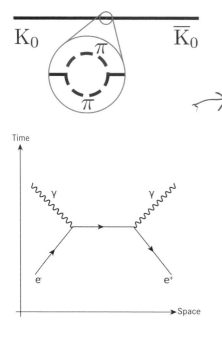

In the late 1960s, a young professor at Columbia University named Edward Tryon attended a seminar given by one of the leading cosmologists of the time, Dennis Sciama. During a break, Tryon suggested that perhaps the universe is a vacuum fluctuation. The young physicist was speaking in earnest, but Sciama took it as a joke. Yet what was heard in that room at Columbia University was the birth of the first scientific idea that attempted to answer the riddle of the origin of the universe, what might have happened moments before the Big Bang.

Sciama's laughter made Tryon abandon his idea until 1973, when he published an article in the journal *Nature* entitled "Is the universe a vacuum fluctuation?" His central argument was that all the energy of the universe, including the mass of all objects it contains, is exactly offset by its gravitational energy and so its sum is, by definition, negative. That is, the sum of all energy in the universe is zero and that allowed the universe to arise literally from nothing. And the most important point of all: this creation *ex nihilo* does not violate any laws of physics.

According to quantum mechanics, the vacuum is not really empty but full of so-called virtual particles and antiparticles, which are created and destroyed at random. In a microscopic region an electron and a positron can arise suddenly, and then almost immediately disappear in too short a time to be detected: such a process is called quantum fluctuation. What Tryon was trying to say that day is that the entire universe emerged in this way. He himself summed up his approach perfectly with these words: "The universe is one of those things that happens from time to time."

Edward Tryon is a professor of Physics at Hunter College in New York City, specializing in theoretical models of quarks, general relativity and cosmology.

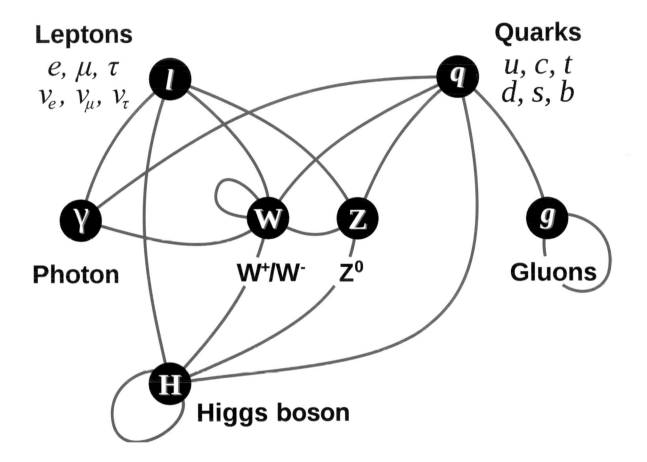

Leptons
$e,\ \mu,\ \tau$
$\nu_e,\ \nu_\mu,\ \nu_\tau$

Quarks
$u,\ c,\ t$
$d,\ s,\ b$

l

q

γ

W

Z

g

Photon

W⁺/W⁻

Z⁰

Gluons

H

Higgs boson

THE OLDEST ORGANISM IN THE BIOSPHERE

A group of scientists has discovered, in the Mediterranean Sea, the oldest known organism: the *Posidonia oceanica*, an endangered species endemic to this region that can live up to 200,000 years. In 2006, in the Balearic Islands, a *Posidonia oceanica* plant was discovered which was estimated to have an age of 100,000 years. It was found in a seagrass meadow measuring 270 square miles (700 sq km). Until that discovery, a bush in Tasmania that exceeded 43,000 years held the record.

Throughout the Mediterranean Sea, large tracts of seagrass meadows can be found, formed by a single genetic lineage, which means that all plants are actually clones of the first that settled. The seagrass shares this clonal growth with the rest of the marine angiosperms (flowering plants).

This process involves the continuous division of the meristems (regions where new cells are produced) and rhizomes. On the one hand, small fragments of the plant, generally stem portions, can result in complete plants as with the cuttings of many garden plants. Moreover, from its rhizome, an underground stem which accumulates food and grows horizontally, parallel to the ground, stems emerge which allow the plant to grow over a much larger area. Generally, these rhizomes end up splitting and separating, but the plants that have already formed are maintained, and are actually clones of each other. The growth process is slow and it is estimated that, to occupy the current area, the first specimen would have to be at least 12,500 years old. More realistic estimates stand at around 200,000 years.

The *Posidonia oceanica* is a fundamental element in the Mediterranean ecosystem, as it performs water purification, feeds numerous species, and has an important task in protecting the coastline against erosion. The authors of the study have taken the opportunity to point out that seagrass is in decline. The increasing water temperature and acidity is affecting this species, as well as boat propellers, particularly for recreational use, which cut up these ancient seagrass meadows as if they were lawn.

The oldest living being in the planet is the species Posidonia oceanica, *an essential marine plant in the ecosystems of the Mediterranean Sea.*

FOUCAULT'S PENDULUM

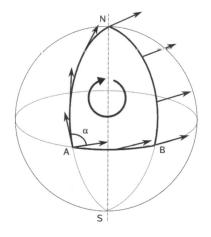

Jean Bernard Léon Foucault (1819–1868) was a French physicist who orchestrated one of the most spectacular experiments in the history of science; using a pendulum, he showed that the Earth turns on its own axis.

After carrying out some tests in his workshop at the Observatoire de Paris on March 26, 1851, during the time of the World Exhibition, Foucault made a public demonstration at the Pantheon in Paris. He suspended an iron cannon ball, weighing 62 pounds (28 kg), from the dome of the Pantheon using a 219-foot (67 m) steel cable. The period of this pendulum (the round trip duration of the swinging ball) was 16 seconds.

To start the movement, the pendulum was separated from its vertical position and held steady with a rope. The rope was set alight, and when it had burned down enough, it broke, allowing the pendulum to swing. This avoided enforcing any momentum in any direction other than its natural swing.

Beneath the point of suspension, there was a circle of damp sand with a radius of about 10 feet (3 m), on which a metal needle placed on the bottom of the sphere drew a path. The pendulum was constructed without a device to compensate for the loss of energy by friction with air, so it needed a new pulse every 5 or 6 hours. Within minutes, the needle track had thickened and, within hours, the scanning sector was more than 60 degrees, that is, the oscillation plane had turned at that angle.

Depending on the latitude of the plane on which it oscillates, the time it takes to make a total turn varies. If the pendulum were installed at the North Pole, the plane of oscillation would rotate at the same speed as Earth, that is, it would take 24 hours to complete a full rotation. In contrast, at the Equator the oscillation of the ball would be infinite, as it does not experience rotation. The Parisian Pantheon pendulum rotates about 270 degrees per day and takes 32 hours to complete the circle.

The Foucault pendulum still remains installed at the Pantheon in Paris, and its operation can be seen in many museums of science and at the United Nations building in New York.

THE RING THAT CROSSES THE MILKY WAY

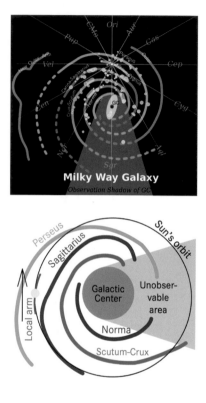

Milky Way Galaxy
Observation Shadow of GC

A team of astronomers has found that the mathematical symbol of infinity is drawn in the stars, in the center of the Milky Way. From the depths of space, the *Herschel* infra-red space telescope has obtained images of a formation of gas and interstellar dust that distinctly forms the shape of the infinity symbol.

Previously, astronomers had only managed to glimpse a part of this ring that lies 600 light years away. The orbiting telescope *Herschel*—created by the European Space Agency in collaboration with NASA and launched in May 2009—has captured the ring for the first time in its full extent. In the images, a ring composed of cold, dense gas mixed with cosmic dust can be clearly seen. This is an area of new star formation, and the images captured by the telescope suggest that the temperature of the bands are only 15 degrees Kelvin, that is, 496 degrees below zero on the Fahrenheit scale.

Thanks to the observations of the *Nobeyama* telescope in Japan, we know that the ring of gas moves as a whole, at a constant speed with respect to the galaxy. The ring is in the center of the so-called Milky Way bar, a region rich in stars in the middle of the arms of the spiral galaxy. The bar is in turn within a larger ring.

The scientific community has not yet described in detail the formation of bars and rings in spiral galaxies, but computer simulations show how such structures could arise as a result of gravitational interactions. For example, the center bar of the Milky Way might have been influenced by the Andromeda galaxy, a neighbor of ours.

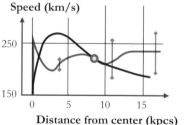

The orbiting telescope Herschel *is named after the German astronomer Friedrich Wilhelm Herschel (1738–1822), discoverer of Uranus and a large number of celestial bodies.*

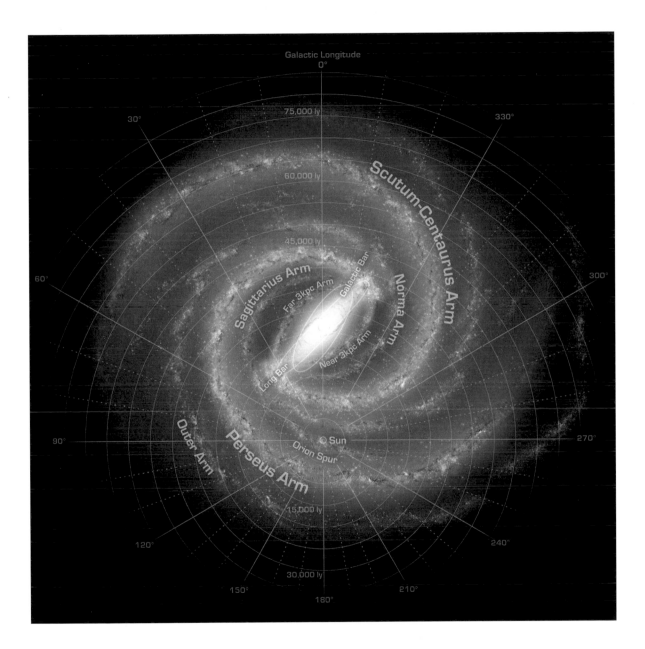

Galactic Longitude
0°

30°   330°

75,000 ly

60,000 ly   Scutum-Centaurus Arm

45,000 ly

60°   Sagittarius Arm   300°

Far 3kpc Arm   Galactic Bar   Norma Arm

Long Bar   Near 3kpc Arm

90°   ⊙ Sun   270°

Outer Arm   Orion Spur

Perseus Arm

15,000 ly

120°   240°

30,000 ly

150°   210°

180°

THE IMMORTAL JELLYFISH

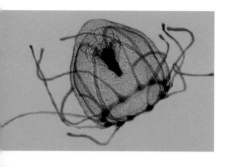

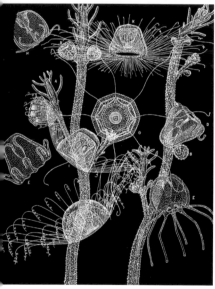

Unlike all the other animals, the *Turritopsis nutricula* does not die within a certain period after reaching the adult stage: it reprograms its cells and becomes a juvenile specimen, a polyp, which in turn will mature again. This is the equivalent to a butterfly that could return to being a caterpillar.

During adulthood, if the water temperature lowers or if there is a shortage of food, the *Turritopsis*, with all its matured and differentiated cells, drops to the seabed as if it were dead. But instead of dying, the jellyfish transforms: its organs and muscles disappear, and in a few hours its whole body melts until it is no more than a bunch of undifferentiated cells. Then this shapeless mass will reorganize and begin to grow anew, developing into a new polyp.

The most remarkable part of the regression of the jellyfish into a polyp is the change that the cells undergo during the process. Normally each cell has a precise function and form, but the cells of this jellyfish are capable of losing their specific function and re-acquiring an almost embryonic capacity to engender new types of cells. A tiny jellyfish cell can become a nerve cell in the new polyp, following a process that defies the laws of biology: changing the cells after they have been differentiated, making them revert to previous stages before their specialization. This phenomenon is called transdifferentiation, and can be observed in animals that can regenerate certain organs or tissues, such as salamanders or starfish. Yet the *Turritopsis nutricula* is the only known form of life able to regenerate its entire body over and over again.

Laboratory studies showed that 100 percent of the specimens had matured and rejuvenated dozens of times without losing their characteristics or abilities in those changes. Researchers had to conclude that organic death is something that in this species simply does not apply. Despite this ability, most *Turritopsis* jellyfish fall victim to the usual threats of life in the plankton world, such as being eaten or disease.

The existence of this incredible creature has been known for decades. In recent years, it has been subjected to original genetic and biological studies to attempt to learn the secret of its immortality.

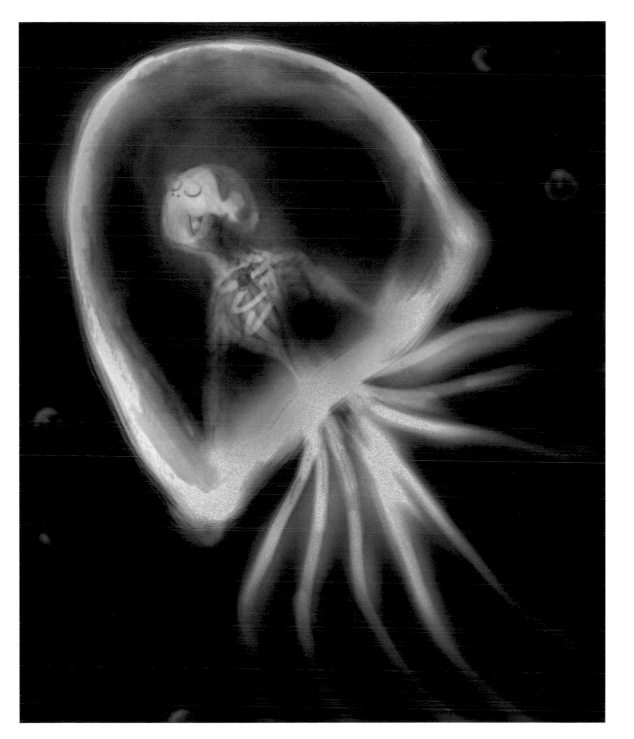

COPY

DARK MATTER, INFINITE PROPULSION

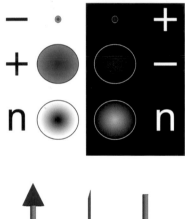

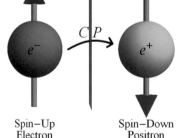

Spin–Up
Electron

Spin–Down
Positron

The entirety of outer space is extremely vast. For example, the closest star to our solar system, Proxima Centauri, belonging to the star system Alpha Centauri, lies about 4.2 unattainable light years away, more than 200,000 times the distance from the Earth to the Sun, or 50 million round trips to the moon. At present, these mind-boggling distances mean that the stars are out of the reach of human exploration. One need only consider that the fastest interstellar spacecraft built to date, *Voyager 1*, which is now leaving the Solar System at an approximate speed of 10 miles (16 km) per second, would need 74,000 years to reach Proxima Centauri.

We would need a spacecraft capable of crossing the cosmos at the speed of light for humans to reach the stars in our lifetime. Different proposals have been suggested, such as vehicles powered by hydrogen bomb blasts, or by the annihilation of matter and antimatter, and even large sailboats with giant reflective sails driven by laser beams.

Unfortunately, all these ideas have their drawbacks and it's doubtful that they can cover such distances, but new possibilities to reach the stars are being studied. Specifically, we refer to that provided by the physicist Jia Liu from the University of New York, who designed a spacecraft powered by dark matter.

Using magnetic fields generated by the spacecraft itself, it would be driven by forcing the accumulated hydrogen gas to pass through nuclear fusion energy and expel the products that would generate impulse, and since dark matter is so abundant in the universe, Liu imagines a rocket that would capture its fuel on the way, converting its autonomy into infinity. According to most models, dark matter particles are their own antiparticles, so when particles collide they decay producing secondary particles of baryonic matter that can be directed to the rear of the vehicle, producing a propulsive force in response. According to best estimates, in a couple of days the ship would reach the speed of light.

Jia Liu drew inspiration for the idea of his spacecraft from the project by the American physicist Robert Bussard, which was based on magnetic fields generated by the rocket to capture the thin gas of interstellar space.

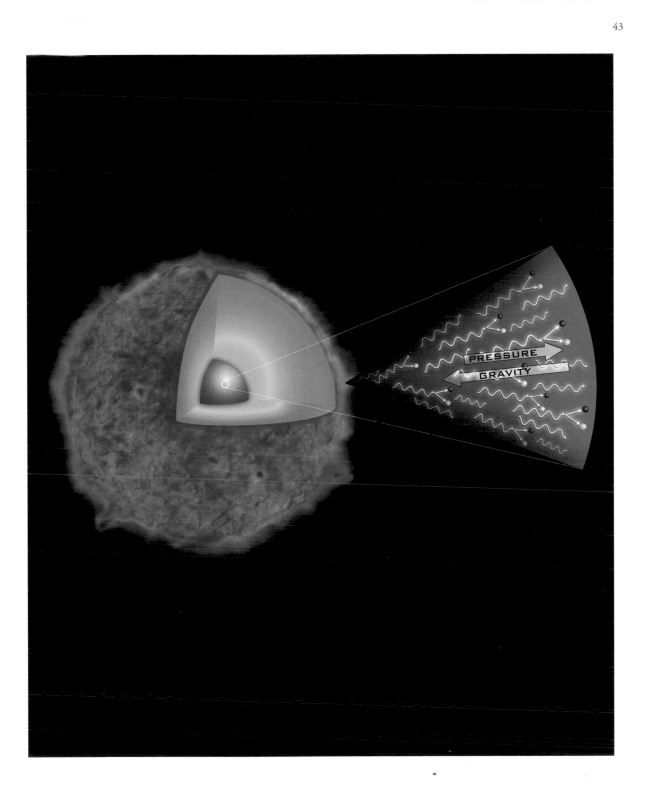

APEIROPHOBIA

This word describes a persistent, abnormal or unjustified sense of fear of infinity or vast and indefinite things. *Apeiron* is a Greek term that refers to the vastness, to indefinite matter, to infinity. The word "phobia" comes from *Phobos* (in Greek mythology, the personification of fear, son of Ares and Aphrodite), and in psychology means an emotional health disorder characterized by an intense, disproportionate fear to objects or situations. The phobia of open spaces (agoraphobia) and phobia of enclosed spaces (claustrophobia) are two of the best known. Phobias are classified within the general chapter of the anxiety disorders, because this is the predominant symptom in this type of alteration.

It is possible to develop a phobia over virtually anything, and there are many rare phobias such as the one presented here. Apeirophobia is a fear that appears when the subject faces infinity or immensity, for example while contemplating the immensity of the universe. Individuals suffering from apeirophobia like having their boundaries and distances defined, and appreciate all that is predictable.

A person suffering from a phobia tends to avoid the phobic situation and recognizes that their fear is excessive or unreasonable, but cannot control it. Exposure to the phobic stimulus often causes physical discomfort: uncontrollable tremors, dizziness, excessive sweating, palpitations, etc. In the most extreme cases, the reaction may present as a panic attack.

Most people with phobias realize that they suffer from a disproportionate or irrational fear, but this recognition does not stop them from feeling an intense emotional reaction to what horrifies them.

THE DIRAC SEA

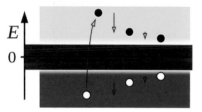

Paul Dirac (1902–1984) was a physicist who made fundamental contributions to the initial development of quantum mechanics and quantum electrodynamics. Among other discoveries, he formulated the Dirac equation, which describes the behavior of fermions, and predicted the existence of antimatter. Dirac shared the Nobel Prize in Physics with Erwin Schrödinger in 1933 for his contributions to atomic theory.

The Dirac Sea is a concept born from the Dirac equations formulated in his work on quantum mechanics. The result of these equations was unusual, since they could predict the existence of particles with negative energy. These particles are known today as virtual particles (the term *virtual* is applied as opposed to *real*).

Dirac hypothesized that there are negative energy states and that space—in our reference system, the energy of a point in the void is zero—was just the total of these complete negative energy states. Moreover, as the total energy of these negative states is smaller than the mass at rest of these quanta—according to the theory of relativity, $E = mass\ at\ rest \times speed\ of\ light^2$ (positive and therefore always *Energy > 0*)—these become purely virtual and no longer exist, and that is why "there is nothing" in a void.

Therefore, the Dirac Sea is the totality of virtual quanta that "fills" the void, in the same way that drops of water fill the oceans.

With respect to virtual particles, sometimes they exist for a brief span of time. This phenomenon occurs when for some fluctuation of energy they happen to have enough energy to "generate" their mass at rest. However, during all these events the law of conservation of charge and the law of conservation of energy must be respected. Due to these conservation laws, Dirac was able to predict the existence of the positron (the antiparticle of the electron) years before its detection in experiments.

Before its experimental discovery in 1932, the positron, the antiparticle of the electron, was conceived as a hollow in the Dirac Sea.

THE INFINITE UNIVERSE OF ISAAC NEWTON

PHILOSOPHIÆ
NATURALIS
PRINCIPIA
MATHEMATICA.

Autore *JS. NEWTON*, Trin. Coll. Cantab. Soc. Matheseos Professore Lucasiano, & Societatis Regalis Sodali.

IMPRIMATUR·
S. PEPYS, *Reg. Soc.* PRÆSES.
Julii 5. 1686.

LONDINI,

Jussu *Societatis Regiæ* ac Typis *Josephi Streater.* Prostat apud plures Bibliopolas. *Anno* MDCLXXXVII.

Sir Isaac Newton (1642–1727) was a physicist, inventor and mathematician, author of the *Principia*, in which he described the law of universal gravitation and established the foundations of classical mechanics through the laws that bear his name. His other discoveries include work on the nature of light and optics and the development of mathematical calculation. Newton was the first to demonstrate that the natural laws that govern motion on Earth and those that govern the motion of all celestial bodies are the same. He is often described as the greatest scientist of all time, and his work is considered the culmination of the Scientific Revolution.

The *Principia* was published in 1687, and it unified, under an elegant mathematical structure, the discoveries of terrestrial mechanics by Galileo with Kepler's description of planetary motion. This synthesis successfully managed to explain the movement of both the objects that surround us in our daily lives and the planets in their orbits. The incredible predictions of the theory, such as the frequency of Halley's Comet, elevated Newton's mechanics into the definitive description of physical reality. In Newton's universe, celestial movements were the result of the attractive force of gravity. His law of universal gravitation along with his three laws of motion formed a large-scale explanation of the known universe.

However, in 1692 Richard Bentley informed Newton of a paradox that could not be ignored: since the gravitational force is always attractive, a finite universe made of stars would inevitably collapse upon itself, falling on its center of masses and forming a spherical mass. Newton countered that the universe should be infinite and matter should be evenly or uniformly distributed in it. Thus, any part would be attracted to the whole in all directions with equal intensity and all the forces would cancel out each other. That is, the universe should have no center of masses.

According to Newton, "the infinite" consists of the intersection of three infinite elements: absolute space, absolute time and matter. It is necessary to define the universe as infinite in both the principle of mechanics and that of universal gravitation.

THE SPACE-TIME DIMENSION

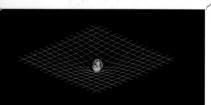

The space-time dimension is the geometric entity in which all physical events in the universe occur. According to Einstein and his theory of relativity (proposed in 1905), space and time are not separate concepts, but are closely linked. As the universe has three observable spatial physical dimensions (height, length, depth), it follows deductively that we consider time as the fourth dimension and space-time as a four-dimensional space to emphasize the inevitability of considering time as another geometric dimension. Thus, we speak of the "space-time continuum."

Time, within the concept of infinite space, acquires a relative character, that is, it depends on the actual geometry of space-time, which is affected by the presence of matter. In fact, the theory of general relativity predicts a universe in which gravitational attraction arises from the curvature caused by mass in space-time. Similarly, Einstein posited that time as we know and appreciate it does not exist, but varies depending on the conditions to which it is subjected: time goes faster or slower depending on the speed and severity of the object.

Many years passed until this effect was demonstrated, and we now know that time passes faster if we move vertically, and slower if we move horizontally. The most famous test took place in 1971 when a group of scientists put atomic clocks on different planes and flew around the world. The results showed that a slight difference of 184 nanoseconds was recorded between the time of the clocks that had remained on Earth and those aboard the aircraft, thus time had passed more slowly for the clock in motion.

From this it was deduced that if we could travel at the speed of light, time would stop and, seemingly, time travel would be possible. If time was also two-dimensional, we would rotate in time the same as we do in space. This would allow us to travel to the past. If so, it would fundamentally alter our experience of cause and effect, among other phenomena.

The theory of relativity not only solved many of the problems that had concerned scientists up to the time of its proposal, but it also dared to contradict some ideas of Isaac Newton that until then were untouchable.

THE INFINITE COOPERATION

The endosymbiotic theory proposed by Lynn Margulis (1938–2011) transformed the Darwinian model of evolution, including the tenets regarding its mechanisms.

Margulis presents a reality where there is no room for anthropocentrism. According to her view of evolution, all bodies seem to cooperate to be able to survive, where individual differences do not generate competitive advantages but, contrary to what Darwin thought, evolution and survival depend on the quality of their union and not on the supremacy of a few.

Life as we know it can be divided into two groups: the prokaryotes (bacteria or archaea, without nucleus) and eukaryotes (cells with nucleus, constitutive of all multicellular organisms). The criterion by which Margulis conceived this division is the fundamental difference between the groups' use of resources.

Some 3,500 million years ago, the Earth was very different from how we know it today and it would have been impossible for mammals to live in such conditions. However, this habitat was conducive to open the path for life, and so the first bacteria began to form in the oceans, and later they altered the entire planet by generating oxygen, which was released into the atmosphere, fixing atmospheric nitrogen in the soil and in the water and preparing the ground, little by little, for another form of life. This was possible thanks to cooperation, as it happened that a prokaryotic organism entered an eukaryotic cell, and this gave both new possibilities, as well as the ability to survive.

This was only the beginning of an infinite number of clusters of cells that would gradually generate more complex organisms, capable of colonizing other habitats and tapping new resources, provided through cooperation and endosymbiosis.

Lynn Margulis was considered one of the most prominent biologists of the last decades and a substantial authority in the field of evolution.

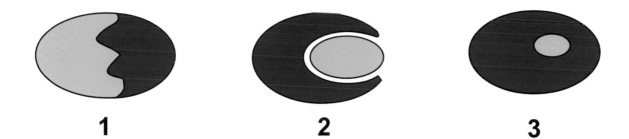

1   **2**   **3**

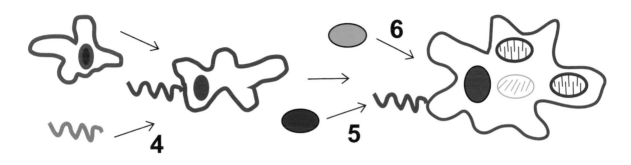

6

4

5

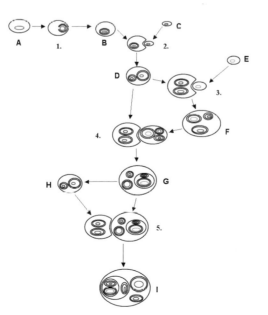

SCIENCE   **THE CHAOS THEORY**

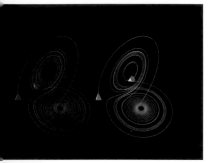

Science has always aimed to reach comprehensive explanations, to find relations between events and express them in a mathematical language that allows us to understand and predict nature. However, there exists an inherent chaos in nature preventing predictions from being accurate. To understand this chaotic nature a new discipline arose, called the science or theory of chaos, which provides a method to find order where before there was only random, erratic, unpredictable events, that is, chaos. In the words of mathematician Douglas Hofstaedter, "it turns out that an eerie type of chaos can lurk just behind a facade of order—and yet, deep inside the chaos lurks an even eerier type of order."

Unlike phenomena concerned with the theory of relativity and quantum mechanics, chaos theory reveals knowledge of everyday life, such as cloud formation or growth of ice crystals. These seemingly disorganized processes have certain quantifiable characteristics: their development in time depends very sensitively on how variables are distributed at the instant when you start observing the phenomenon in question.

Edward Lorenz (1917-2008), one of the fathers of chaos theory, studied how to predict the weather, and for this purpose he developed a highly simplified mathematical model consisting of 12 equations. One day in 1961, he wanted to check some data and re-entered the numbers into the computer, but to save time, he did the calculations with 3 decimal places instead of 6.

According to conventional wisdom, the results should have been altered only very slightly. However, from the slight changes in the values of the initial variables there arose widely divergent solutions. Lorenz described this phenomenon as the "butterfly effect."

The butterfly effect is used to describe phenomena in which tiny changes in the variables used cause large-scale changes in the final results.

PLANCK UNITS

In 1899, physicist Max Planck (1858–1947), founder of quantum theory and Nobel Prize winner for Physics in 1918, proposed units of length, mass, time and temperature from the fundamental constants of nature: the gravitational constant G, the speed of light in a vacuum c and the Planck constant h. This resulted in what he called the system of natural units, as they are based on constants of nature and not arbitrary standards without physical basis such as the meter, kilogram and second.

Planck length is defined as the distance below which the laws of classical physics are no longer satisfied due to the emergence of quantum gravity effects, so it makes no sense to speak of motion and, therefore, time. It is equivalent to the distance light travels in Planck time, which corresponds to a size one billion billions smaller than a proton radius.

$$\ell_P = \sqrt{\frac{\hbar G}{c^3}} \approx 1.616252\,(81)\times10^{-35}\,\text{m}$$

Planck time is determined by measuring how long it takes light to travel Planck length. From the perspective of quantum mechanics, it is considered that Planck time represents the smallest unit which could be measured, that is, it would not be possible to observe any difference between the universe in a specific instant and any separate instant for less than one Planck time. However, images taken by the *Hubble* space telescope have cast doubt on this theory.

$$t_P = \sqrt{\frac{\hbar G}{c^5}} \approx 5.39124\times10^{-44}\ \text{seconds}$$

Planck mass is the amount of mass (21.7644 micrograms) that, included in a sphere whose radius was equal to Planck length, would generate a density of 1093 g/cm³. According to current physics, this would have been the density of the universe when it was about 10^{-44} seconds old, the so-called Planck time.

$$M_p = \sqrt{\frac{\hbar c}{G}} = 2.18\times10^{-8}\,\text{kg}$$

Planck temperature stands out from the rest because it provides a fundamental limit of quantum mechanics, since it is defined as the maximum value of temperature: the so-called infinite temperature. The Planck temperature is the temperature of the universe during the first moment, that is, the temperature during the first unit of Planck time in the Big Bang. The fundamental unit for Planck temperature would be equivalent to 3.5 x 10^{32} degrees Celsius.

$$T_P = \frac{m_P c^2}{k} = \sqrt{\frac{\hbar c^5}{G k^2}} = 1.4167\times10^{32}\,\text{K}$$

According to current estimates, the Big Bang took 2 units of Planck time to create all the forces of the universe.

WHITE HOLES

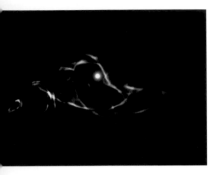

The existence of black holes begs a basic question: where does all the stuff that it devours go? This question is the basis of the theory about the existence of white holes. Black holes are based on the theory of Einstein's relativity, a symmetrical theory that therefore implies that there should be an "exit," or something like the "other side" of the famous wormhole.

Unlike black holes, for which there is a well studied physical process called gravitational collapse (which gives rise to black holes when stars more massive than the sun run out of nuclear "fuel"), there is no clear analogous process that would explain the generation of white holes.

If white holes exist, they would represent a finite area of space-time with a density which would give them the ability to warp space but, unlike the black hole, they would expel matter and energy instead of absorbing it. In fact, no object can remain within the region for an infinite time, that is why a white hole is defined as the time reverse of a black hole.

Another widespread theory is that white holes would be very unstable, would last a very short time and, actually, after their generation, they could collapse and become black holes, which would limit us when studying them.

Presently we must wait for a new phenomenon that can help to confirm or rule out the existence of white holes. And we should bear in mind that until very recently, the existence of black holes was also in question.

Little is known about white holes, beyond that they are the opposite of black holes, that is they would launch matter at the universe rather than attract it.

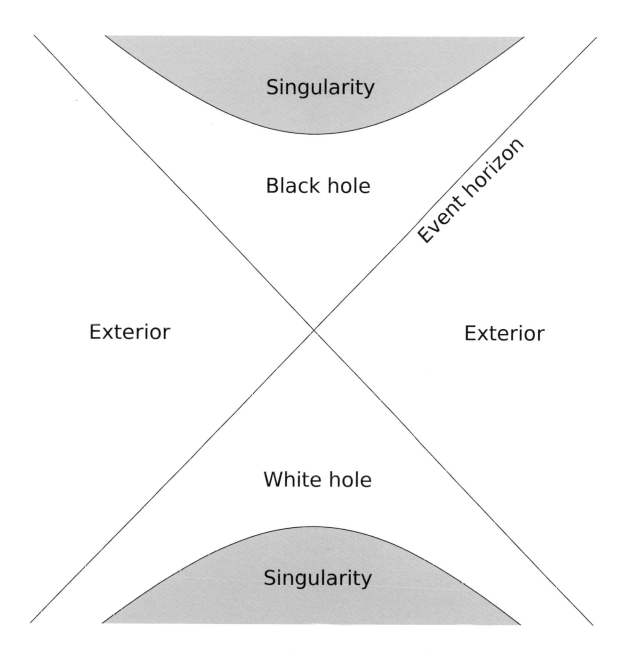

HALLEY'S COMET

The appearances of Halley's Comet have been documented since the year 240 BC, by civilizations ranging from the Chinese and Koreans, to the Indians and Japanese. This comet is undoubtedly the one that has attracted the most interest, because it is very bright and in its appearances it comes very close to our planet.

Halley's Comet travels around the Sun on a very elliptical orbit. Its distance to the perihelion (the point of its orbit at which it is closest to the Sun) is 0.6 AU, between the orbits of Mercury and Venus, while its aphelion (farthest from the Sun) is 35.3 AU, nearly the distance of the orbit of Pluto. Curiously, its orbit is retrograde, going in the opposite direction of other planets.

The comet was named after the English astronomer Edmund Halley (1656–1742), who discovered its periodicity in 1705. According to Newtonian mechanics, it was understood that a comet entered the Solar System, orbited around the Sun following a parabolic orbit, and set off to infinity, thus it could only be observed once. However, Halley conducted a study of comets from ancient references and calculated their orbits as accurately as he could. In his observations, Edmund Halley found that the characteristics of the comet described in 1682 coincided with those of 1531 (described by Petrus Apianus) and 1607 (observed by Johannes Kepler): they all shared the orbital inclination angle and the closest distance to the sun.

With this data, it was concluded that they were not three different comets, but three instances of the same comet that moved in an elliptical orbit with a period of 75–76 years. Thus it was deduced that comets, when they moved farther away from the Earth, do not head to infinity but rather follow an orbital path that would eventually bring them back. A new appearance of the comet in 1759 confirmed this theory, and since then, every 76 years, humanity awaits the visit of this large, bright comet. We saw it happen in April 1986 and expect a return visit in 2061.

Halley's Comet has its origin in the Oort Cloud, which is situated on the edge of the Solar System, almost 1 light year from the Sun. Although it originally was considered a long cycle comet, the gravitational pull of the gas giants has shortened its orbit, trapping it within the Solar System.

ZERO-POINT ENERGY

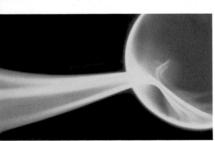

The concept of zero-point energy was proposed in 1913 by Albert Einstein and Otto Stern. It derives from quantum mechanics, the science that describes the behavior of particles of atomic dimensions. It is defined as the lowest energy a system can possess, or in other words, it is the residual energy of a system once all possible energy has been extracted. By definition, zero-point energy cannot be extracted or used. If we can extract more energy, we have not reached the zero point.

The funny thing is that the zero-point energy is not zero. To understand this without the need for complicated equations, we will consider the duality of matter: all elementary particles are particles and waves. All matter we see around us is made up of particles which are also waves. A wave vibrates, swings, and undulates, that is it has a frequency of oscillation. Every oscillator has an energy associated with this frequency—which is like the kinetic energy due to the motion of oscillation. The only way to measure at null energy is to be at frequency zero, that is, the point of oscillation cessation. But a wave that does not oscillate is not a wave, and particles are waves.

Another way to tackle this issue is from the point of view of Heisenberg's uncertainty principle. This principle states that we cannot precisely and simultaneously determine the position and velocity of a particle: the more certain the position, the more uncertain the velocity, and vice versa. Thus, we may have zero uncertainty in one of the two magnitudes, but in this case, we will have infinite uncertainty in the other. This uncertainty is negligible in the macroscopic world, but in the subatomic world it is of great relevance.

The concept of zero-point energy and the desire to extract plentiful, sustainable energy from the vacuum has attracted the attention of inventors of all ages, as outlined by many perpetual motion machines. The operation of these machines would violate the second law of thermodynamics, and so they are a seemingly impossible object.

THE SUPERFLUIDITY OF HELIUM

Superfluidity is a state of matter characterized by complete absence of viscosity, what makes it different from a *very fluid* substance, which would have a viscosity close to zero, but not exactly zero. Thus, a superfluid element in a closed circuit would flow endlessly and indefinitely without friction. It was discovered in 1937 by Pyotr Kapitsa, and independently by John F. Allen and Don Misener, and its study is known as quantum hydrodynamics.

Superfluidity is defined as a physical phenomenon that occurs at very low temperatures, near absolute zero (0 K or -273 °C), a limit in which all activity ceases. At these temperatures, almost all elements freeze, with the sole exception of helium, which is liquefied at ambient pressure at a temperature of 4.2 K (-269 °C). At that point of extreme cold, its viscosity (resistance) becomes zero. Experiments with helium isotopes have shown that, in a state of superfluidity, a substance is capable of passing through solid surfaces such as glass, pushing through the microscopic pores of a glass and passing through it like a sieve.

This ability of superfluids to cross any solid object appears to be attributable to its strong capacity for oscillation, which reflects the quantum hydrodynamic model, and it can be understood by considering that in the absence of viscosity and friction, the constant movement of the molecules manages to open the path through the particles of the elements, which it passes. In experiments with these superfluids, we can see the ability of eternal movement of these substances, which are capable of climbing the walls of containers and emptying them, forming infinite fountains. These fountains are produced when there is a helium flow between two different pressure zones, which creates a jet that flows from the high pressure zone to the low pressure zone. The peculiarity of these ghostly fountains is that they can flow eternally if the conditions of the experiment remain constant.

Helium is a noble gas with unusual properties. It is the product of hydrogen fusion, and is widely found in the Sun and in oil wells in the Earth.

THE QUANTUM VACUUM

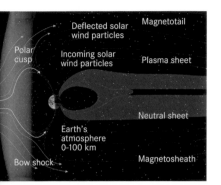

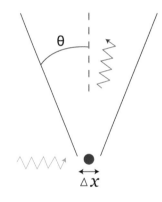

For centuries, *horror vacui* was a principle presumed (by scientists and philosophers) to dictate far-reaching consequences in the natural world: it was believed that nature abhorred a vacuum and tended to fill it with something, whether it be air or "ether." Today, it is common to accept that space may be empty. However, when in today's physics it is theorized that the phenomenon of particle emission from a vacuum is common, or it is postulated that the emission of particles from a black hole is made by the surrounding empty space, without a doubt we enter into the realm of the paradox: the fact that a vacuum, which is by definition the absence of matter, emits matter is actually the less surprising.

The origin of the idea lies in Heisenberg's uncertainty principle, which establishes a limit on how accurately you can measure simultaneously the position and speed of a particle. An equivalent of this principle imposes a similar restriction on the possibility of measuring the energy of a particle and the moment when measurements are made.

By combining this principle with the equivalence between energy and mass proposed by Einstein, we come up with the surprising result that matter can be created from a vacuum. A particle can suddenly appear and disappear immediately. This cannot happen according to classical physics—matter cannot be created out of nothing—but it is possible according to quantum physics, provided that the lifetime of the particle is sufficiently short. Indeed, a particle of mass m has an energy mc^2, and if its lifetime is less than h/mc^2, according to the uncertainty principle, the particle cannot be detected: during that time, the mass (energy) of the particle is below the margin of error with which it could be measured.

Thus, according to quantum mechanics, a vacuum is full of particles that appear and disappear, concealed by Heisenberg's uncertainty principle. Such particles, undetectable by definition, are called virtual particles. The presence of virtual particles in a vacuum leads to a series of physics problems that have not yet been resolved. The fundamental difficulty is that the vacuum energy is formally infinite, because you can create virtual particles with unlimited energies.

One of the biggest challenges for theoretical physicists today is resolving the inherent difficulties within the concept of the quantum vacuum. In particular, the possibility of microscopic black holes that evaporate quickly complicates the concept of the quantum vacuum even more: as virtual particles are formed, virtual black holes can be created from the vacuum.

Werner Heisenberg (1901–1976) formulated the uncertainty principle, a foundation for the development of quantum theory. This principle states that it is impossible to simultaneously determine the position and speed of a particle accurately.

AT-C-G-A-G-C

DEOXYRIBONUCLEIC ACID, AN INFINITE HELIX

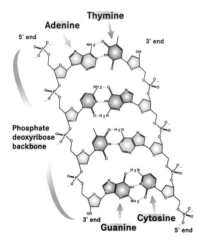

DNA or deoxyribonucleic acid was first described in 1869 by the Swiss physician Friedrich Miescher (1844–1895), who observed a very abundant substance in body fluids which he called nuclein. Almost a century and a half later it has been established that DNA is the fundamental basis of life and evolution.

DNA is a nucleic acid consisting of a series of four different nucleotides: adenine (A), guanine (G), cytosine (C) and thymine (T). Depending on how these nucleotides are joined together (e.g., A-A-T-G-T-G, C-C-T-G-A-T-G, and other combinations), they will form a sequence which can be interpreted as a language and shared by all species that inhabit the planet, whether trees, bacteria or human beings, and which is passed on from generation to generation, allowing the inheritance of characteristics that identify each individual and each species.

These DNA sequences are what we now call genes, and are what defines the difference between a dragonfly and an elephant or between an ant and a human being. The major structural features of DNA, depending on the species, can be in the form of a single molecule, a single chain which is joined at its ends to form a circle (as in the case of bacteria), or several molecules which bind in complementary and antiparallel pairs, forming a double helix which may remain more or less stretched, or very compact, as in the case of chromosomes.

Perhaps the most astonishing feature of DNA is that it is an infinite molecule: in order to pass from generation to generation, it duplicates itself in a process called "semiconservative," that is, each individual of the offspring will inherit a parent molecule and a newly formed molecule, so that the new organism will always be part of the predecessor's organism, allowing the DNA to perpetuate in subsequent generations.

In 1953, researchers James Watson and Francis Crick described for the first time the famous double helix structure or spiral staircase, the DNA model which we currently use.

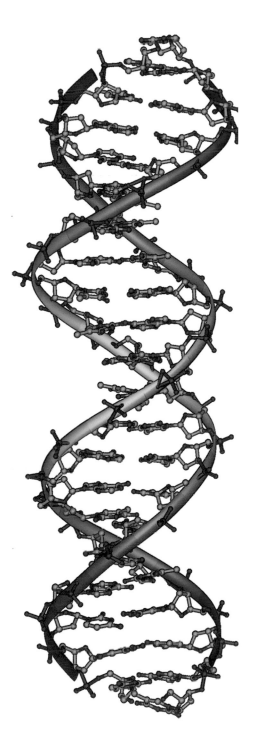

BLACK HOLES

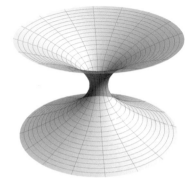

A black hole is a celestial body with such an incredibly strong gravitational field that not even electromagnetic radiation or light can escape it. Its mass is concentrated at a point of infinite density that is called a singularity. At that point, gravity is an almost infinite force, such that it actually modifies space-time. As we move away from the singularity, the gravitational influence decreases, reaching a point where the speed needed to escape the black hole equals the speed of light.

The hole is limited in space-time by the so-called "event horizon." The event horizon separates the black hole region from the rest of the universe, and it is a boundary space surface from which no particle can escape, including light.

The formation of a black hole occurs because of a phenomenon known as gravitational collapse, discovered in the mid-20th century by scientists such as Robert Oppenheimer and Stephen Hawking. This process begins in the moments after the death or total extinction of the energy of a red giant. Given the mass of the star, its gravitational force begins to exert a force on itself so powerful that it is able to concentrate all its mass in a very small volume, resulting in a white dwarf. This phase of the process, which can last for billions of years, ends with the final collapse of the body of the star by its own gravitational attraction, which then converts into a black hole.

In other words, a black hole is the result of the force of extreme gravity pushed to the limit. The same gravity that keeps the star stable compresses it to the point where the atoms begin to collapse. Electrons in orbit become closer and closer to the atomic nucleus and eventually merge with protons to form more neutrons, culminating in a neutron star. At this point, gravity grows exponentially with decreasing distance between the atoms. Neutron particles implode, flattening even more and creating a black hole: infinite gravity in an immeasurably small space.

As early as 1796, the mathematician Pierre Laplace (1749–1827) suggested the idea of an object with a mass concentration that could even trap light.

STRING THEORY

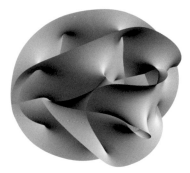

String theory is a theoretical model conceived in the last decades of the 20th century; the theory centers on small particles of energy that vibrate like violin strings, producing a symphony unique to each particle of the universe.

In order to understand string theory we must go back to the problem that triggered it. For over half a century, the laws of physics were subject to two very reliable and irrefutable theories each in its scale, but, when combined, they presented serious inconsistencies and anomalies.

The theory of relativity explains how the force of gravity works, and that model works very well for large bodies such as galaxies and planets. In contrast, subatomic elements behave according to the theory of quantum mechanics, responding to three fundamental forces: the strong nuclear force that holds together the protons and electrons, the weak nuclear force, responsible for radioactive decay, and the electromagnetic force. The standard model describes the behavior of these particles and forces with pinpoint accuracy, but with one notable exception: gravity, which is very difficult to describe microscopically.

While the relationship between the four fundamental forces and matter explains all the events of the universe, problems arise when we try to reconcile the chaotic behavior of quantum mechanics with the harmonious balance of the theory of general relativity. For many years this has been one of the greatest challenges of theoretical physics: formulating a quantum theory of gravity, to be known as the theory of everything or unified theory. And string theory seems to have achieved that.

Until recently, scientists have described the basic components of matter (atoms and subatomic particles) as small spheres or zero-dimensional points. Yet string theory states that matter is composed of vibrating loop-shaped threads of energy called strings at the subatomic level. The strings vibrate in certain ways, providing the particles with their unique properties, such as mass and load. Depending on how they vibrate, we see a photon, a quark or any other particle of the standard model.

String theory represents a total revolution in theoretical physics, and there are many detractors of this final unification model, because there are many parts that cannot be demonstrated empirically, such as massless subatomic particles (gravitons), faster-than-light particles, or the need to accept that the universe is contained in eleven dimensions, rather than four as believed (three spatial and one temporal), for the theory to be valid.

String theory is also extremely complex, so much so that it contains mathematical questions that are beyond the capacity of our current knowledge. All these questions have become a new challenge for modern physics.

THE INFINITE MICROBIOTA

When we think of organisms able to conquer and colonize different environments, we tend to place human beings at the top of the list, but the reality is very different: we are no more than mere passing visitors in a world completely governed by an intractable number of microorganisms, among which the most abundant are bacteria.

Bacteria have always been the most abundant organisms on the planet, and since their inception, 160 million years ago, they have been responsible for giving the world the features that have made life possible at more complex levels.

Bacteria are the only beings capable of living in every ecological niche, as they have multiple mechanisms of genetic recombination that give them the capacity for rapid mutation and adaptation to new habitats within an incredibly short period, while any other animal organism would require years. Bacteria have enabled oxygen to accumulate in the atmosphere and nitrogen in the soil, allowing other forms of life to take advantage of these conditions; they have decomposed rock into fertile land, providing ecosystems with the organisms and elements necessary to perpetuate life. The human body, composed of 100 trillion cells, is colonized by a microbiota of 90 billion organisms of 200 different types that live in perpetual balance and without which our lives would be endangered.

The number of bacteria in the universe is incalculable, but it is clear that it verges upon infinity.

The term "microbiota" refers to the community of living organisms in a certain niche, the most commonly found include Staphylococcus aureus, Escherichia coli *and* Candida albicans.

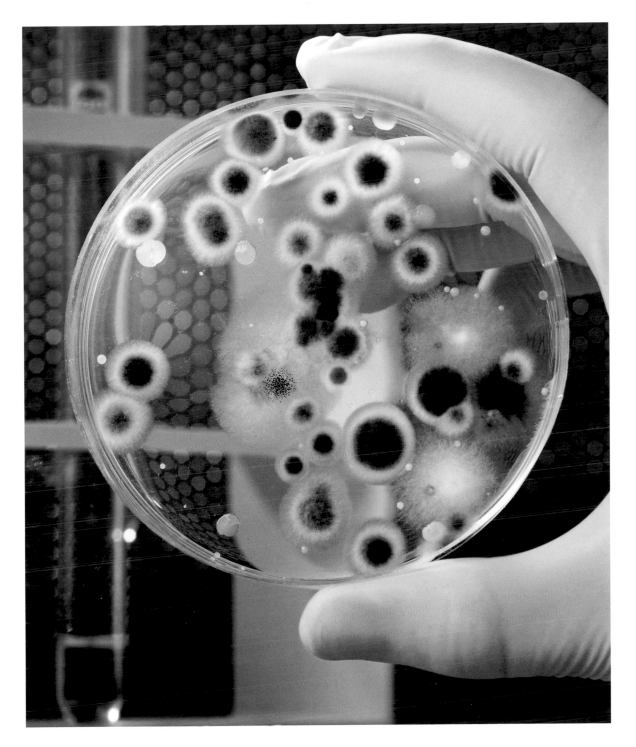

SCIENCE

THE ULTRACENTRIFUGE

A centrifuge is a device present in most laboratories that causes chemical samples to rotate at high speed until the components are separated by molecular weight. The ultracentrifuge is a very powerful model: the first ultracentrifuge created forces of up to 5,000 times the force of gravity, and current models are capable of generating an acceleration that exceeds even a million g (9,800 km/s^2).

We owe the appearance of the ultracentrifuge in 1923 (announced in 1925) to the Swedish chemist Theodor Svedberg (1884–1971), who received the Nobel Prize in Chemistry in 1926 for his research based on ultracentrifugation as a method of protein separation. Later in the 1970s in the UK, a record centrifugation speed of 7,250 km/s^2 was reached, three times higher than that of a supersonic aircraft.

Currently, ultra-centrifuges are indispensable in the fields of molecular biology, biochemistry and the use and manipulation of polymers in general. Their use is so prevalent in laboratories that we've become inured to their incredible capacity. Currently, new technologies have pushed the speed of rotation to such extremes, reducing friction by implementing vacuum conditions and temperature control, so that, if we look at some models, we will discover that instead of the full speed appearing, the symbol of infinity appears in the speed indicator.

In addition to inventing ultracentrifugation, Theodor Svedberg carried out influential work on a vaccine against polio.

DIVISIBILITY OF THE ATOM

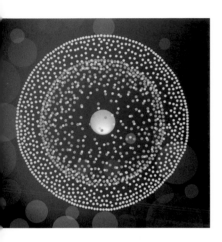

The British scientist Joseph John Thomson (1856–1940) established in 1897 that the atom was divisible, and presented a model in which the atom could be represented as a positive sphere in which electrons are embedded.

In the mid-19th century, studies of the electrical nature of matter and, in particular, the spark that flies between two charged objects that are near, brought into question that the atom is an indivisible particle. The best evidence came during the final years of the century and proved that the atom was divisible and also had an electrical nature. These conclusions emerged when it was observed that the fluorescence in the wall of a tube was produced by an invisible radiation emerging from the cathode, which they called cathode ray.

In 1897, Thomson established that the cathode ray consisted of beams of negatively charged particles, and that the components of the cathode rays were not charged atoms, but smaller particles resulting from the fragmentation of the atom. These identical, negatively charged particles that are present in the atoms of all elements were thus named electrons.

From this hypothesis, Thomson proposed an atomic model in which the atom, a positively charged sphere, is surrounded by negatively charged electrons uniformly distributed, whose number should be sufficient to neutralize the positive charge that the element contains.

From this discovery it was deduced that the atom, as well as electrons, must contain in its interior other possibly positive particles, which were called protons and which were attributed (incorrectly, as it would turn out) to the total size of the mass of the atom's nucleus.

Thomson's atomic model is also known as "raisin bread." In 1908, Ernest Rutherford, a former student of Thomson, proved that the raisin bread model was inaccurate.

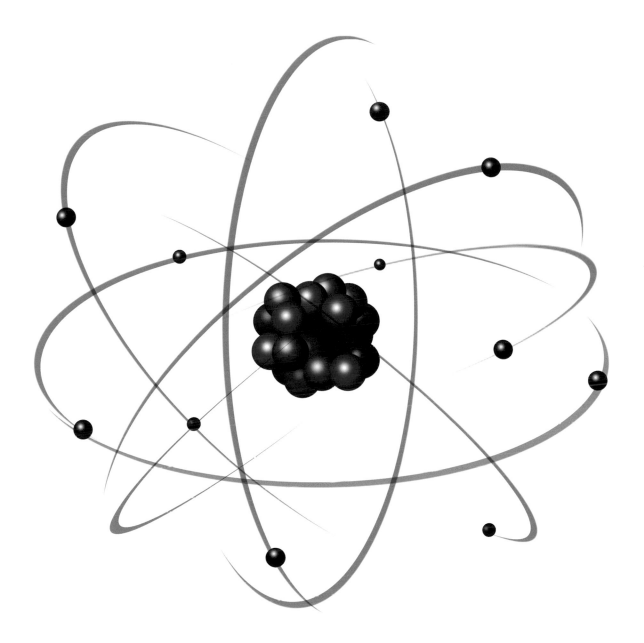

QUANTUM TUNNELING

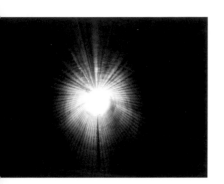

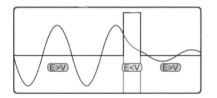

The development of quantum mechanics has allowed the study of the behavior of particles at a subatomic scale, which was impossible to understand with classical physics, given that nanoparticles show an erratic and confusing behavior and experience effects that, on the large scale, escape all the laws of traditional physics. One of the most interesting subatomic phenomena is the tunnel effect, which would be equivalent, on the large scale, to a ball passing through a wall.

In short, quantum tunneling means that an electron (or quantum particle) enters and passes through an area which, in principle, would be prohibited. When we say that the area is prohibited for the electron, we mean that the electron does not have enough kinetic energy (which is due to its speed, in terms of a classic analog) to pass through that area, because there is a "barrier" that should prevent it from passing through.

Classical physics dictates that the total energy is the sum of kinetic and potential energy, therefore, the energy is always equal to or greater than the potential energy. The case in which total energy is less than the potential one, from a classical physics point of view, represents unreachable states to a particle. So, the point at which the total energy equals the potential represents a "turning point": the particle cannot advance, it must go back, like a ball that hits a wall.

However, in quantum mechanics something different happens: at subatomic scales, a particle may have a total energy lower than the potential energy and is able to cross the turning point like a tunnel, violating the principles of classical physics and, depending on the difference between its energy and the potential value, to penetrate a larger or smaller distance. In this way, it is possible that an electron reaches a barrier, passes through it, and appears at the other side, although it should be noted that the tunnel effect does not always occur: as with all quantum effects, it is a matter of probabilities.

Quantum tunneling was described in 1928 by physicist and astronomer George Gamow. This phenomenon also appears in other branches of physics, and in electronics there are transistors that base part of their operation on this effect.

RUTHERFORD-BOHR ATOMIC MODEL

1.7×10^{-5} Å

1.1 Å

By 1900, physicist Ernest Rutherford's (1871–1937) experiments showed that matter is not as strong as previously believed, and that actually for the most part it is spectral. Rutherford proposed that the atom consisted of a nucleus with all the positive charge and most of its atomic mass concentrated, and tiny particles orbiting like planets around the sun. These particles, the electrons, would have a negative charge and would spin at such a speed that their centrifugal force would compensate for the force of electrostatic attraction exerted by the core.

The number of positive charges in the core would equal the number of orbiting electrons, so that the electric charge of the core and that of the electrons would neutralize each other. The positive charge of the center of the atom explained the stability of matter in space, while the electrons orbiting freely explained the ability of the electrons to pass from one atom to another, enabling electrical current.

Rutherford's planetary model did not specify the speeds of the electrons or their distances from the atomic nucleus. Yet it was a concept that the physicist Niels Bohr (1885-1962) took up some time later.

Bohr's model initially focused on the hydrogen atom, and owing to its simplicity, it is still used as a simplification of the structure of matter. The hydrogen atom according to Bohr's model has a nucleus with one proton and one electron revolving in the first orbit around the nucleus, the one with the lowest energy. According to the postulates of Bohr's atomic model, electrons are arranged in circular orbits that determine various energy levels whose number is limited, unlike the proposal in the Rutherford model, accepting an infinite number of orbits. When the electron spins in these orbits, it does not emit energy. If energy is transmitted to the electron, it will fly from the first orbit to another with a higher energy, and when it returns to the first orbit it will emit energy in the form of light radiation.

Ernest Rutherford was a physicist and chemist who was devoted to the study of radioactive particles and the disintegration process of the elements. He was a professor of Niels Bohr, the physicist who made major contributions to atomic structure and quantum mechanics.

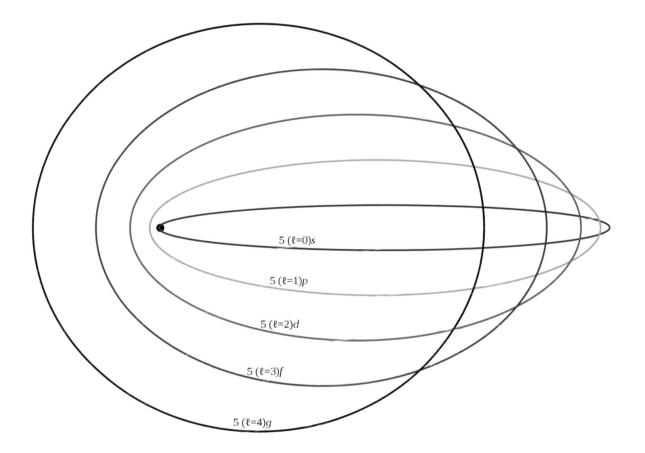

5 (ℓ=0)s

5 (ℓ=1)p

5 (ℓ=2)d

5 (ℓ=3)f

5 (ℓ=4)g

_NO_

MENDEL'S LAWS

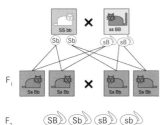

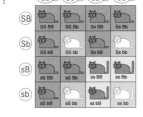

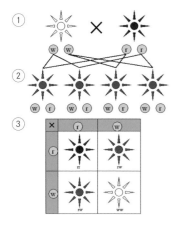

Gregor Mendel (1822–1884), considered the father of genetics, was a monk whose experiments on the transmission of hereditary characteristics have become the foundation of the current theory of heredity. Mendel's laws explain the traits of the offspring from knowledge of the characteristics of their parents, and the laws have made it possible to understand the phenomenon of evolution and the emergence of the endless features and characteristics that enable the diversity of individuals and species.

Through the work that he began in 1856 experimenting with crosses of different species of peas in the monastery garden, he was able to infer the proportions in which hereditary traits are expressed and in this way describe, without the use of any other tool other than his own observation, the principles that later became known as Mendelian inheritance laws or Mendel's laws.

Mendel was able to distinguish the dominant characteristics (those that are always expressed) from the recessive (which are expressed less often), and was even able to determine the exact frequency of recombination and predict the number of individuals who presented either trait. Another pillar that he established was the difference between genotype (the information in the genes) and phenotype (the characteristics represented in the genes that were expressed).

Each organism has two hereditary factors for each of its characteristics: one inherited from one parent and the second from the other. Mendel studied the color of the pea seed (yellow and green), two antagonistic or different characteristics for the same trait. He crossed the seeds and achieved a generation in which all seeds were equal. By crossing these plants, he obtained a generation with the following ratio: ³⁄₄ with yellow seeds and ¹⁄₄ with green, that is, a 3:1 ratio.

After studying how antagonistic characteristics are inherited, Mendel studied the non-antagonistic traits such as shape and color of the seed. To do this, he crossed two pure breeds, smooth yellow and rough green, and he observed the first and second generations of descendants, the latter with a 9:3:3:1 phenotypic ratio. Mendel concluded that each non-antagonistic heredity factor is inherited irrespective of the others, and is grouped randomly in descendants.

Gregor Mendel's work went unnoticed by the scientific community until 1900, when the Mendelian laws were rediscovered by three botanists: Hugo de Vries, Eric von Tschermak and Karl Correns.

DARK MATTER

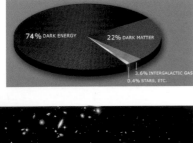

Let's consider the vastness of planets, stars and other celestial bodies that make up the many galaxies in the universe. Surprisingly, the mass of these imposing objects, known as ordinary matter, only represents 4 percent of the total mass of the universe, while there is an invisible ingredient that is still unknown and represents about 23 percent: the dark matter, the thread with which the infinite fabric of the universe is woven. Its existence is essential so that the stars in a galaxy remain grouped and so that star formations remain associated in clusters.

Identifying the nature of this ubiquitous and mysterious mass that is present throughout the universe, as well as identifying dark energy (which makes up the remaining 73 percent of the total mass), has become a great challenge for modern astrophysics. Although dark matter cannot be seen or directly detected since it does not emit or absorb electromagnetic radiation (light, radio waves, etc.), its presence is manifested through the force of gravity it exerts on other celestial bodies, as it is capable of curving the path of a light beam so that the images that come from other galaxies appear slightly distorted. The interpretation of the many images collected over the past decades has enabled astronomers to calculate the total mass of the cluster of dark matter that is attributed to the distortion, even when this mass is invisible.

There are several theories to explain the composition of this dark matter. First, dark matter could be brown dwarfs, with masses around $1/20$ that of the Sun and which by failing to reach a temperature sufficient for the combustion of hydrogen in its core, are not bright enough to be detected directly. A second possibility is that dark matter is made up of supermassive black holes that could be at the center of many galaxies. In both cases, it would be baryonic dark matter, that is, composed of protons and neutrons, such as ordinary matter. The third possibility is that dark matter is a form of matter yet to be discovered, consisting of unknown non-baryonic elementary particles that would have formed shortly after the Big Bang.

The existence of dark matter was proposed by the physicist and astronomer Fritz Zwicky (1898–1974). His work was continued by astronomer Vera Rubin, who has been able to provide the most direct proof of its existence.

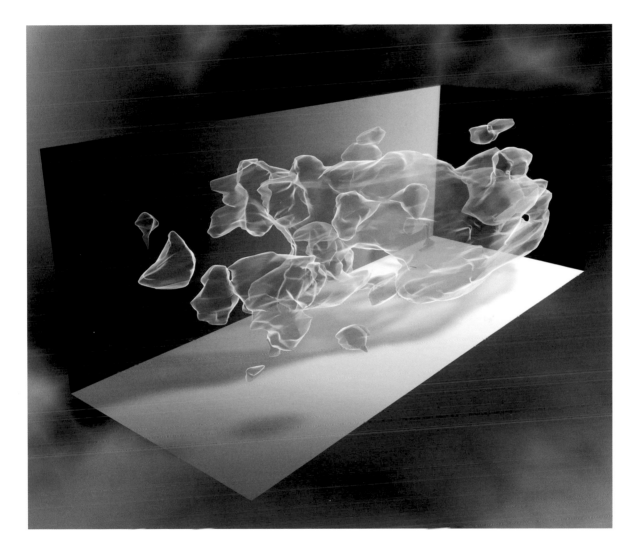

CHARLES DARWIN AND EVOLUTION

Gibbon   Human   Chimpanzee   Gorilla   Orangutan

Charles Darwin (1809–1882, opposite) laid the foundations of evolutionary theory by suggesting that all forms of life have evolved from a common origin and by a process of natural selection. At that time, geologists defended the catastrophic theory, according to which life responded to a succession of individual creations and each species was immutable, so it did not undergo any change over time.

At 22, Darwin was admitted to a scientific expedition around the world by boat. The voyage on board the *Beagle* lasted 5 years and profoundly influenced the young naturalist, who returned to England determined to devote his life to science.

From his observations, Darwin discovered that closely related species existed that differed in structure and eating habits, and he concluded that these species were not always so, but that various circumstances had caused them to develop structural and functional modifications. He also concluded that adaptation to different environments generated changes in organisms, which explained the diversity of life on the planet.

Thus arose the idea of common descent and the concept of natural selection, whereby organisms have a tendency to change; when among the infinite possibilities, there is one trait or characteristic that is an evolutionary advantage, the carrier of this change will be favored, increasing their ability to leave a descendant. Thus, the offspring that will inherit this characteristic will be a new line with its own features, which will make it prevail over the other species.

In 1837, Darwin began the first book of notes on the origin of species. On November 24, 1859 *On the Origin of Species by Means of Natural Selection, or the Preservation of Favored Races in the Struggle for Life* came to light, and the first edition of 1,250 copies sold out that very day. The theological implications of the work, which attributed to natural selection powers that until then were reserved for God, caused fierce opposition by some sectors.

The original title of his revolutionary work was On the Origin of Species by Means of Natural Selection, or the Preservation of Favoured Races in the Struggle for Life, *but after its sixth edition, it was shortened to* The Origin of Species.

THE IMMORTAL LIFE OF HENRIETTA LACKS

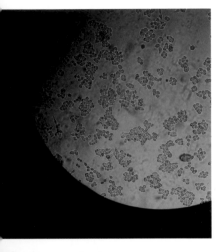

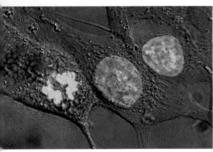

The name Henrietta Lacks is very popular among the scientific community for her unique contribution to science. She was an American woman born in 1920, who was diagnosed with a tumor in the uterus that proved to be malignant and ended her life at the tender age of 31 years. Her doctor, George Gey, took a tissue sample from the tumor and developed the first continuous culture of human cells.

Thus, the first immortal cell line in history, known as HeLa, was formed. The progeny cells are used routinely in laboratories around the world and have contributed significantly to the development of numerous investigations because of their immortality, which derives from their ability to multiply indefinitely as long as they are in the appropriate environment. Thanks to Henrietta's cell cultures a polio vaccine was developed; they are also used in the research for a cure for leukemia and cancer, to analyze the behavior of cells and the growth of viruses, to synthesize proteins, and in genetic studies.

For a long time the eternal life of these cells was a mystery. Outside the human body, cells begin to die slowly and inexorably before they complete 50 divisions. Cells cannot survive without the support provided by the body and cannot live longer artificially because they age, therefore, one way or another they die. In contrast, HeLa cells continue feeding, creating waste and reproducing themselves indefinitely even within a test tube. They are immortal not only because they live outside the human body, but also because they do not age.

Researchers suspect that their aggressive growth and resistance may be due to a combination of a viral infection caused by human papilloma virus and a mutation suffered by the patient that generates a defective protein, P53, which is considered essential in the cellular life cycle.

The popularity of HeLa cells took off from 1975 and it came to the attention of her relatives and descendants, who were unaware until then of the impressive contribution that Henrietta had made to science.

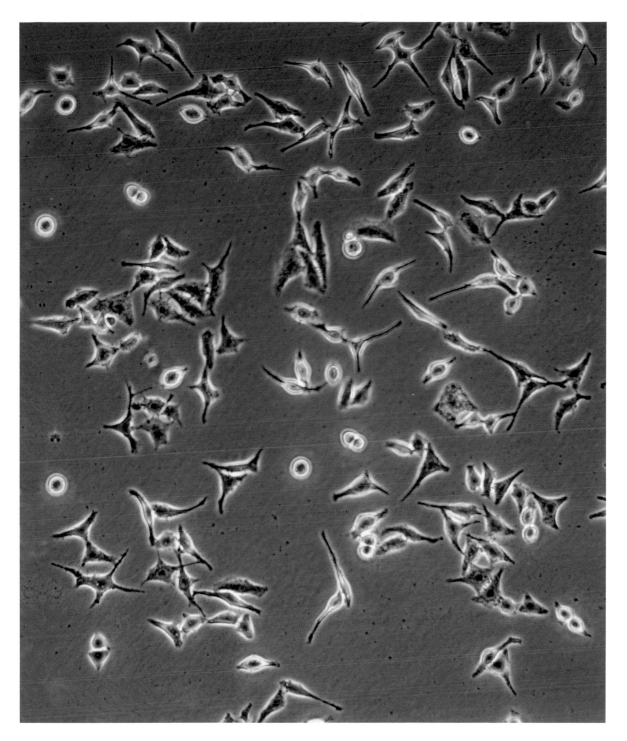

WAVES

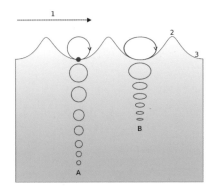

The vast mass of water forming the oceans and seas is subjected to an endless wave motion caused by wind action. The wind exerts a thrust on the sea's surface creating small waves, which in turn offer more resistance to the wind, which causes increasingly larger undulations.

Once formed, the wave no longer depends on the wind, but its own gravity: a wave falls in the valley of the previous wave and the wave spreads almost without losing energy, as it does not move a mass of water. If the wind speed increases, the elevations are higher, increasing the distance between the crests and the propagation speed. When the wind rises, the distance between crests is shortened and the fronts become steeper. After the wind raises the sea surface, gravity pulls the water down, so kinetic energy accumulates with all the rising and falling. At a certain height, the wave does not hold and breaks because it cannot maintain its own body of water. At that time, any accumulated kinetic energy is transformed into water transport.

Until recently, it was believed that "rogue waves" up to 10 feet (30 m) high were a legend, but there is evidence that, although they are rare, they occur in all the seemingly quiet open seas and oceans of the world. Another unique type of wave are tsunamis, which are not related to wind but to earthquakes and volcanic eruptions. In these situations, in the center of the disturbance the water may sink or be lifted explosively. In both cases the movement causes a single wave of frightening dimensions, which moves at speeds of up to 620 miles per hour (1,000 km/h) and reaches heights of over 65 feet (20 m). The Pacific Ocean is the area most affected by this phenomenon.

Using the Douglas Sea Scale it is possible to classify the different states of the sea in 10 degrees ranging from 0 to 9, with reference to the height of the waves. The English Admiral Percy Douglas developed this scale in 1917, when he headed the newly established Naval Weather Service.

THE INDESTRUCTIBILITY OF MATTER

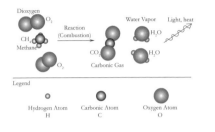

Legend

Hydrogen Atom Carbonic Atom Oxygen Atom
H   C   O

The law of conservation of matter is one of the fundamental laws in all the natural sciences and was formulated by Antoine Lavoisier (1743–1794), considered the father of modern chemistry. Lavoisier studied the chemical reactions and found that the mass (amount of matter) is permanent and indestructible, something that is preserved despite any changes.

Combustion, one of the major problems of 18th-century chemistry, caught the attention of Lavoisier while he was working on an essay on improving the techniques of street lighting in Paris. The chemist found that heating metals like tin and lead in closed containers with a limited amount of air resulted in the metals being covered with a layer of calcine at a specific time during the warming. Lavoisier showed that the calcination of a metal was not the result of the loss of mysterious "phlogiston" (a substance imagined to eminate from burning materials), but the gain of a portion of air.

In 1774, Antoine Lavoisier conducted an experiment by heating a closed glass vessel containing a sample of tin and air. He found that the mass before heating (glass container + tin + air) and after heating (glass container + heated tin + the remainder of air) was the same. Subsequent experiments showed that the product of the reaction, tin oxide, was the original tin with a part of the air. With these tests, Lavoisier observed that oxygen is essential for combustion and formulated the law of conservation of matter: the total mass of the substances present after a chemical reaction is the same as the total mass of the substances before the reaction. This law is often summarized as follows: matter is neither created nor destroyed, only transformed.

The Greek philosopher Democritus of Abdera (460 BC–370 BC) had already proposed the indestructibility of matter with his atomic theory, contained in his fundamental principle "nothing can arise out of nothing; nothing can be reduced to nothing."

THE CARBON CYCLE

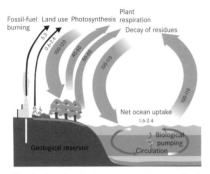

Sea-Surface Gas Exchange (in parts per million)

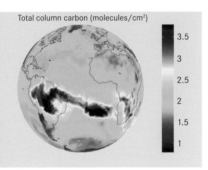

Carbon is a chemical element essential for life that is present in nature in many different forms. All organic molecules—carbohydrates, lipids, proteins, nucleic acids—are formed by linked carbon chains.

The carbon is stored in air, water and soil in the form of a gas called carbon dioxide (CO_2). During photosynthesis, plants consume CO_2 from the atmosphere that they metabolize, so it becomes part of their molecules. When herbivores feed on these plants, they also assimilate the carbon they contain. The animals release much of this carbon in the form of CO_2 during respiration and store the rest in their tissues. Over time, this carbon also returns to the atmosphere by metabolic processes or by the depletion by other animals. During the decomposition of organic matter, bacteria and fungi break down dead plant and animal matter, releasing an amount of CO_2 that can be added to that given off in volcanic activity. The carbon dissolved in the atmosphere is again ready to be captured by plant organisms, which take up the renewal cycle of this element.

In the hydrosphere, the carbon cycle is similar. Aquatic plants use dissolved CO_2 for photosynthesis and marine animals release it into the water by breathing. When the concentration of carbon into the aquatic environment is greater than that in the atmosphere, a carbon exchange is produced between them. Nature is responsible for regulating the concentrations of carbon, removing a certain amount from its natural cycle. This generates the so-called fossil fuels, like oil, coal and natural gas, which are remnants of organic matter that thousands of years ago were buried without oxygen before breaking down, resulting in incomplete decomposition.

The combustion of fossil fuels, especially since the Industrial Revolution, has caused a release of carbon higher than what nature can handle. The result is the well-known greenhouse effect, which exacerbates climate change with consequences as dire as rising sea levels, changes in precipitation and desertification.

Carbon is the most abundant element in the human body (17.5 percent), after hydrogen and oxygen. It constitutes 0.025 percent of the Earth's crust and is abundant in sedimentary rocks, in the deposits of solid and liquid fuels and in the deposits of certain geological layers.

SCIENCE   **THE GAIA HYPOTHESIS**

The Gaia hypothesis postulates that the biosphere, oceans and crust are closely integrated and are in charge of achieving an optimal physical and chemical environment for life. Gaia is a representation of the planet Earth as a macro-organism formed by living organisms that interact allowing the continuity of life. This superbeing is capable of regulating itself through chemical, biological and geological conditions which create suitable conditions that remain relatively constant through actively controlling the global temperature, atmospheric composition and ocean salinity. The theory was devised by the chemist James Lovelock in 1969 and published in 1979, though it was writer William Golding who suggested the name of Gaia, the Greek goddess of the Earth.

Gaia behaves as a system that tends toward equilibrium. If some environmental change threatens life (such as a massive injection of carbon dioxide after a volcanic eruption), she would act to restore equilibrium (more phytoplankton would appear in the oceans to absorb carbon dioxide in the water).

This hypothesis considers that the many forms of life not only affect their environment collectively for favorable conditions, but that it is life itself which regulates and controls the environment. In other words, the current conditions of the Earth are not so because life passively allowed it to do so, but rather life caused them to occur. Before life appeared on the Earth 2,500 million years ago, the atmosphere was dominated by carbon dioxide. Life sprang up to absorb the gas, generating nitrogen (bacteria) and oxygen (photosynthesis).

The global temperature has remained unchanged for millions of years, even though solar radiation has been increasing steadily. Therefore, a continual warming of the Earth should be occurring, but this has not happened. In response to greater solar radiation, carbon dioxide (with its heat retention properties) decreased accordingly, as though Gaia were acting through the world's plants to keep it at the optimum temperature to sustain life.

James Lovelock's inspiration for the Gaia hypothesis emerged while he worked for NASA on the Viking *project, in which a probe was sent to Mars in order to study the possibility of the existence of life on the "Red Planet."*

SCIENCE

THE TARDIGRADE

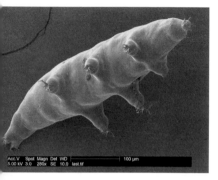

Tardigrades, commonly called water bears, are a family of microscopic invertebrates that live in moist environments. Typically they are found in the film of moisture covering mosses and ferns, although there are also oceanic and freshwater species of tardigrades. These animals range in length between 0.004 and 0.05 inches (0.1–1.5 mm), and their larvae can be smaller than 0.001 inches (0.05 mm). They have no circulatory, respiratory or excretory systems.

The tardigrade has a feature that makes it one of the most amazing species on the planet. These animals are among the few species that have the ability to enter a state of cryptobiosis (reversibly suspend their metabolism) when the environment is not conducive to their needs, and can remain hibernating for hundreds of years. Through a process of dehydration, they can undergo a reduction in body moisture percentage from 85 percent to 3 percent. In this way, growth, reproduction and metabolism of tardigrades is reduced or stopped temporarily, while it awaits better conditions.

This resistance enables tardigrades to survive extreme cold and dry periods, intense radiation, heat and pollution of any kind. Studies have shown that they can live at temperatures close to absolute zero (−273°C or −460°F) or as high as 151°C (304°F). Not surprisingly, then, water bears inhabit every corner of the planet and live in regions from the Himalayan mountains to ocean trenches over 13,123 feet (4,000 m) deep, from the polar regions to the tropics. They can also withstand radiation levels of 1 thousand times higher than other animals.

In the normal life of a tardigrade, it can survive 7 years of dry hibernation several times. Returning to life takes them 10 to 15 minutes, and only advanced tardigrades or those who have not dried symmetrically do not revive. The water bear is not immortal; it is estimated that its life expectancy is over 70 years.

The resistance of water bears passed the toughest of tests in 2007 when they were loaded on the Foton-M3 *spacecraft and exposed to the vacuum at an altitude of 886 feet (270 km). The water bears not only survived the vacuum but on return they reproduced without any complications.*

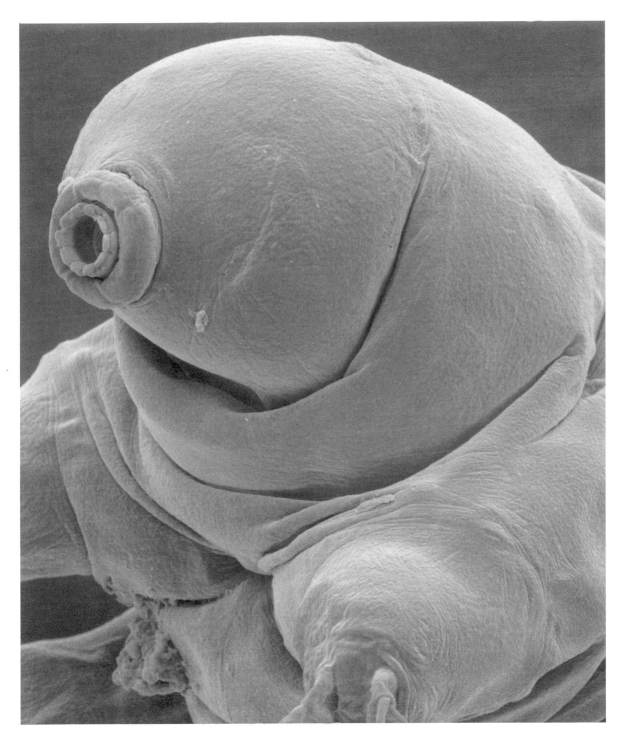

SKIP

COLORS

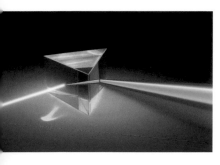

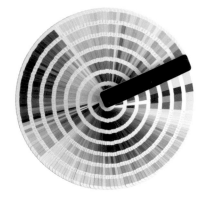

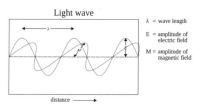

Light wave

λ = wave length

E = amplitude of electric field

M = amplitude of magnetic field

distance ⟶

Nature offers us the spectacle of the decomposition of white light during one of the effects that has most fascinated humanity—the rainbow. The physicist Isaac Newton (1643–1727) studied and reproduced this phenomenon using a prism, and when in 1704 he published his treatise *Opticks* he established that there were seven colors: red, orange, yellow, green, blue, indigo and violet.

According to the theory, every time a ray of light passes through a water droplet suspended in the atmosphere, it refracts, showing all the colors that make up visible light. Each of these colors correspond to different wavelengths of the components in the spectrum of visible light, which are separated as they exit the water drop due to the fact that the angle of refraction differentiates between wavelengths.

According to one interpretation of this theory, the better the white light is separated by refraction, the greater number of colors that are perceived. Although Newton referred to seven colors, there are endless shades, as it is a continuum in which one color turns into another.

The human eye has three types of color receptors called cones, which can associate the primary colors blue, red and green, corresponding to the wavelength to which each cone type is most sensitive. It could be argued therefore that there are only three colors, but this triad of colors is not the generator of the infinity of colors that can be perceived by the eye or used in the arts. Thus, besides the secondary colors, which are obtained from the combination of the primary colors in this initial triad, pastel colors for example cannot be obtained, as they are colors to which white has been added in different proportions.

Newton listed seven colors because he was a follower of the law of seven, which was then believed to govern the universe, as there were seven metals in alchemy (gold, silver, copper, mercury, lead, tin and iron), seven celestial stars (Sun, Moon, Mercury, Venus, Mars, Jupiter and Saturn), seven musical notes and seven days of the week.

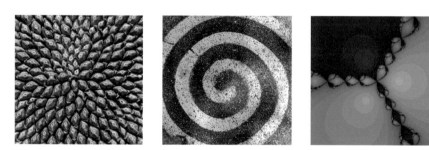

MATHEMATICS

LEIBNIZ'S INFINITESIMAL CALCULUS

Gottfried Leibniz (1646–1716) was a German philosopher, mathematician and states-man. He was one of the great thinkers of the 17th and 18th centuries and is recognized as the "last universal genius." His work addresses not only philosophy and mathematical problems, but also theology, law, politics, history, philology and physics.

Leibniz's contribution to mathematics, also attributed to Newton, was to list the funda-mental principles of infinitesimal calculus, the main problems of which were originally squaring (determination of the length of a curve and the areas and volumes of figures) and tangency (how to trace tangents in curves and surfaces). In modern mathematics, calculus includes the study of limits, derivatives, integrals and infinite series. More specifically, calculus is the study of change, just like geometry is the study of space.

According to Leibniz's notes, on November 11, 1675 he first used integral calculus to find the area under the curve of a function $y = f(x)$. Leibniz introduced several notations used today, such as the integral sign $\int$, which represents an elongated S, from the Latin *summa*, and the letter d to refer to differentials from the Latin *differentia*. This calculus notation is probably his most enduring mathematical legacy, as it is the one still used today. Leibniz did not publish anything about his *Calculus* until 1684.

The last years of his life were occupied by the dispute with Newton over who had first discovered calculus. The debate over the "fatherhood" of infinitesimal calculus was grueling and lasted several years. Mathematicians at that time were divided into two groups, the British supported Newton, and those from the continent supported Leibniz, who spent the rest of his life trying to show that he had not plagiarized Newton's ideas. Subsequent research concluded that each had independently discovered calculus, yet in fact Newton did so first. This dispute had very negative effects for British mathematicians, who chose to ignore the method of Leibniz, which was much superior.

In addition to Leibniz and Newton, the brothers Johann and Jakob Bernoulli also partici-pated to a large extent in the development of calculus.

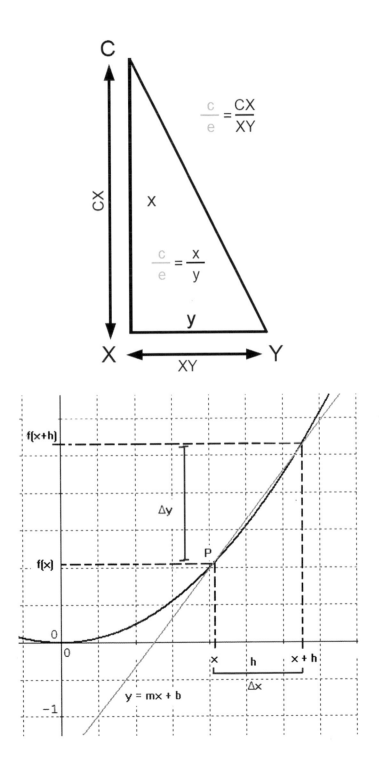

MENGER'S SPONGE

Menger's sponge or cube is one of those curious and surprising mathematical objects known as fractals whose structures repeat shapes in different scales.

Menger's sponge is a structure that taken to infinity would vanish, an object with an infinite surface but enclosing zero volume. Presented in 1926 by the mathematician Karl Menger (1902–1985), the origin of this cube is a three-dimensional composite of Sierpinski's carpet, observed by his colleague Waclaw Sierpinski in 1916.

To form one of these carpets, you need a square divided into 9 equal parts (3 across by 3 down), and the center is removed. The same process is repeated with the remaining 8, and so on, indefinitely. The result forms a surface with holes of different sizes in an area that tends to zero as the number of iterations increases.

The structure of Menger's sponge is based on the three-dimensional representation of the square—the cube. Each side of the cube is divided into 9 squares. This subdivides the cube into 27 smaller cubes, similar to a Rubik's cube. We remove the middle cubes from each face and the middle cube, leaving only 20 cubes. If we repeat this process an infinite number of times we obtain the Menger's cube, also known as sponge due to its resemblance to this marine animal.

Karl Menger received his doctorate in 1924 from the University of Vienna and he was professor of geometry at the same center from 1927 to 1938. He actively participated in the Vienna Circle and made fundamental contributions to the progress of 20th-century mathematics.

MATHEMATICS

NUMBER *e*

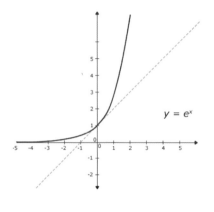

$$y = e^x$$

$$e^{i\pi} + 1 = 0$$

$e = 2{,}71828182845904523536028747135266249775724709369995$
$95749669676277240766303535475945713821785251664274$
$27466391932003059921817413596629043572900334295260$
$59563073813232862794349076323382988075319525101901\ldots$

The number *e*, Napier's constant or Euler number, is an irrational number (with non-periodic infinite decimal places) fundamental in the field of calculus. This constant is the base of natural or Napierian logarithms invented in 1614 by Scotsman John Napier (1550–1617), and published posthumously in 1618, and also the basis of exponential functions. It has been suggested that mathematician Leonhard Euler (1707–1783) gave it the symbol *e* to stand for "exponential." The exponential function of base *e* is the only function whose derivative is itself.

Napier entered the number *e* in some tables referenced in the appendix of a study on logarithms, but a specific value for the number *e* was not given in these tables, just a list of natural logarithms calculated from this new constant. To find the first use of the number *e*, and the first calculation of the first decimals, we have to refer to Leonhard Euler, who first referred to the constant in 1727, and cited it with the letter *e* for the first time in the publication *Mechanics*, in 1737.

The number *e* has many applications in all branches of science and economics. For example, in biology it is applied to calculate exponential growth, which occurs in certain populations of bacteria or in the recovery of a forest area after a fire, and in finance, for something as common as the calculation of bank interest. Similarly, it appears in many technical fields to describe electrical and electronic phenomena.

The approximate value of *e* is:
$e \approx 2.718281824590452354\ldots$

It is believed that the Greek Hippasus discovered irrational numbers trying to write the square root of 2 as a fraction (or ratio), but instead he showed that it cannot be written as a fraction, so it is irrational.

$$e = 2.718\ 281\ 828\ 459\ 045$$

THE FOUR COLOR THEOREM AND
INFINITE COUNTRIES

The four color theorem states that no more than four colors are required to color a map with infinite countries so that two neighboring countries will never have the same color. It is assumed that countries only have one territory, and that the world is round or flat. The precise shape of each country does not matter, the only relevant issue is knowing which country shares a border with another.

Cartographers have known since the Renaissance that four colors were enough so that two neighboring states would not share the same color, but until the 19th century no one thought that this system was related to mathematics, let alone, that there could be a proof to confirm that it was valid for any type of map.

This problem was known at first as "Guthrie's problem," as it was Francis Guthrie (1831–1899), then a student at the University College of London, who first considered it in 1852. Not satisfied with the evidence he was trying to develop, he asked his brother Frederick who presented it to his lecturer, the famous mathematician Augustus De Morgan, who proposed the problem to various colleagues.

The first demonstration appeared in the journal *Nature* on July 17, 1879 and was made by mathematician Alfred Bray Kempe, for which he was awarded Fellow of the Royal Society. Unfortunately, in 1890, Percy John Heawood, an English mathematician who dedicated 60 years of his life to the four color problem, proved that Kempe's demonstration had some errors. Thus, the four color theorem once again assumed the range of conjecture.

In 1976, the mathematicians Kenneth Appel and Wolfgang Haken, with the help of a computer expert, proved the four color theorem using Kempe's method. With over 1,500 possible configurations of maps, and after working for 1,200 hours, they created a computer program that proved that four colors were sufficient to color any map.

It was the first time that the mathematical community accepted a computer-assisted proof, which created much controversy, and so this demonstration and its acceptance generated one of the most massive paradigm shifts in the mathematical world.

The four color theorem states that no more than four colors are required to color a map with infinite countries so that two neighboring countries will never have the same color.

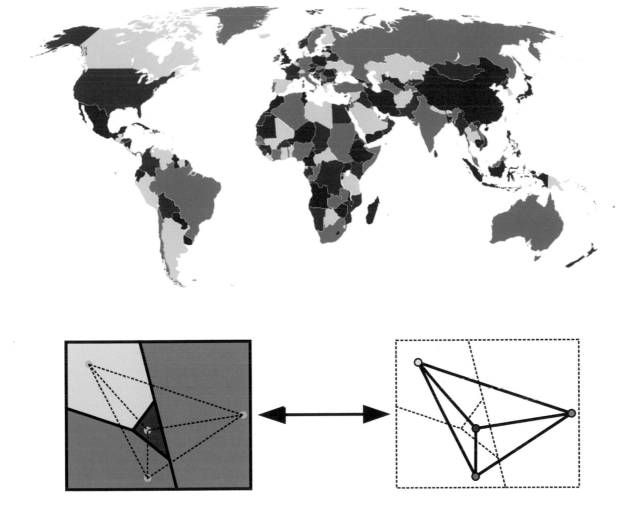

THE INFINITE MONKEY THEOREM

This theorem states that a group of monkeys typing at random on a typewriter for an infinite period of time will almost surely end up writing any given book from the French National Library. In a different way of posing the theory, popular among Anglophones, the monkeys could write the works of William Shakespeare. In this context, the term "almost surely" is a mathematical term with a precise meaning, and the monkey is a metaphor for the creation of a random sequence of letters.

The original idea was formulated by the mathematician Émile Borel (1871–1956) in his article *Mécanique statistique et irréversibilité*, published in 1913. Borel proposed that if a million monkeys typed 10 hours a day, it was extremely unlikely that they could produce something equal to the content of the books of the world's richest libraries and yet, by comparison, it would be even more unlikely that the laws of statistics would be violated, even slightly. For Borel, the purpose of the metaphor of the monkeys was to illustrate the magnitude of an extremely unlikely event.

In the 1970s, it was proposed that the time range imagined be extended to infinity, making it an infinite number of monkeys typing for an infinite time interval. However, insisting on both infinites is excessive—a single immortal monkey that infinitely types on a typewriter could be enough to formulate the theorem.

Surprisingly, there have been several practical experiments that aimed to bring this theory to life. One of the most famous attempts was carried out in 2003 by a group of scientists in Paignton Zoo and the University of Plymouth (England) by placing a computer keyboard in a cage with six crested black macaques for a month. Not only did the monkeys produce five pages consisting of a long series of the letters g, s and q, but the monkeys took to stoning and defecating on the keyboard.

$$\sum_{n=1}^{\infty} P(A_n) = \infty \Rightarrow P(\limsup A_n) = 1$$

Émile Borel was a pioneer of measure theory and its application to the theory of probability. His research on game theory also stands out.

GEORG CANTOR AND THE THEORY OF SETS

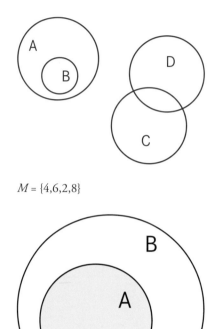

$M = \{4,6,2,8\}$

Georg Cantor (1845–1918) was a German mathematician of Russian origin. He was a founder, along with Dedekind and Frege, of set theory, the basis of modern mathematics. With their research on infinite sets, he was the first to be able to formalize the notion of infinity in the form of transfinite numbers. In 1874, Cantor's first work on set theory appeared. Cantor established the counterintuitive notion that infinite sets do not always have the same size: there are several infinites, some larger than others.

The set we all know is that of natural numbers, consisting of the sequence 1, 2, 3, 4, ... But does this set of natural numbers have a limited number of elements or, conversely, is their amount immeasurable? The first thing that Cantor found is that the set of natural numbers is composed of other sets that have the same number of elements. For example, the set of even numbers has the same amount of elements as the set of whole natural numbers. If we establish a correspondence between the elements of the set of even numbers and the set of natural numbers, we see that for each natural whole number, there is another even whole number. This means, first, that the set of natural numbers is an infinite set, that is, one in which one of its parts is also infinite, since even numbers constitute a set that is part of the set of natural numbers and is infinite.

However, Cantor found that there are sets of numbers that do not correspond one for one with the natural numbers, but on the contrary, they exceed them, so he deduced that some infinities are more infinite than others. To illustrate this discovery, he invented the concept of transfinite numbers, which serve to indicate the degree of infinity in a set. So we know that the infinite has degrees, variations, which in turn are of different infinites in extent, and that infinite sets are infinite parts greater than those that encompass them and beyond. The first letter of the Hebrew alphabet, aleph, is used to indicate the degrees of infinity of infinite sets.

This study of levels or layers of the infinite was described as mathematical madness by his professor Kronecker. The general resistance to Cantor's theory and the vain attempt to establish the continuum hypothesis (unprovable within set theory) burnt the mathematician out, and by 1884 the first signs of mental illness appeared and manifested episodically until Cantor's death at a psychiatric hospital.

Cantor's theorem shows that for every infinite there is a larger infinite, and therefore there is an infinity of infinities.

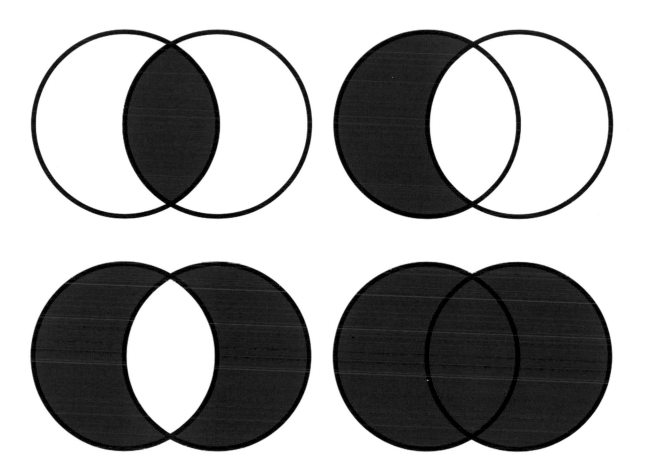

$\phi = 1,6 1...$

$\frac{1+\sqrt{5}}{2}$

THE GOLDEN NUMBER

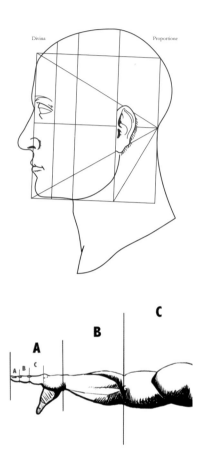

The golden number, also called golden ratio, golden section or divine proportion, is represented by the Greek letter φ (phi) in honor of the sculptor and architect Phidias (c. 490–430 BC). It is an irrational number, that is, with infinite decimal figures that do not have a repeat sequence (which would make it a periodic number). It was discovered in antiquity, not as a unity, but as a relation or proportion between the two straight segments.

Although the golden number did not receive its symbol until the 20th century, its discovery dates back to ancient Greece, where it was well known and used in architectural designs to establish the proportions of the statues and temples, both in their floor plans and in their façades—a clear example is the Parthenon, a work directed by Phidias. Euclid was one of the first to describe this number by the ratio of two lines *a* and *b* which comply with the formula *(a+b)/a = a/b*, which was defined as: "A straight line is said to have been cut into extreme and mean ratio when, as the whole line is to the greater segment, so is the greater to the lesser."

The golden section was much employed in the Renaissance, particularly in visual arts and architecture. It was considered the perfect ratio between the sides of a rectangle, was called a golden rectangle. So too can the golden section can be found in the major works of Leonardo da Vinci. Leonardo's interest in mathematics of art and nature is well known, in fact, in the *Vitruvian Man*, which appeared in 1509 on the cover of Luca Pacioli's work entitled *De Divina Proportione*, it can be seen how all the body parts are related to the golden section, and also how the face of *La Gioconda* features a rectangle which encloses a perfect golden rectangle. Michelangelo also made use of the golden ratio in the impressive sculpture of *David* to determine, for instance, the position of the navel with respect to height, and the placement of the finger joints.

Every day we handle objects in which this number has been considered for their design, for example, most credit cards are a golden rectangle. In nature, there are many elements related to the golden section (or to Fibonacci numbers, with which it is closely linked), such as the arrangement of petals on flowers, the relationship between leaf veins, and even human anatomy, as pointed out by Da Vinci and Michelangelo.

Formula:

$$\phi = \frac{1+\sqrt{5}}{2} \approx 1.6180339887498948482045868343656381177720309...$$

Around 432 BC Phidias carved the statue of Zeus at Olympia, one of the seven wonders of the ancient world, using the proportions of the golden ratio.

Φφ

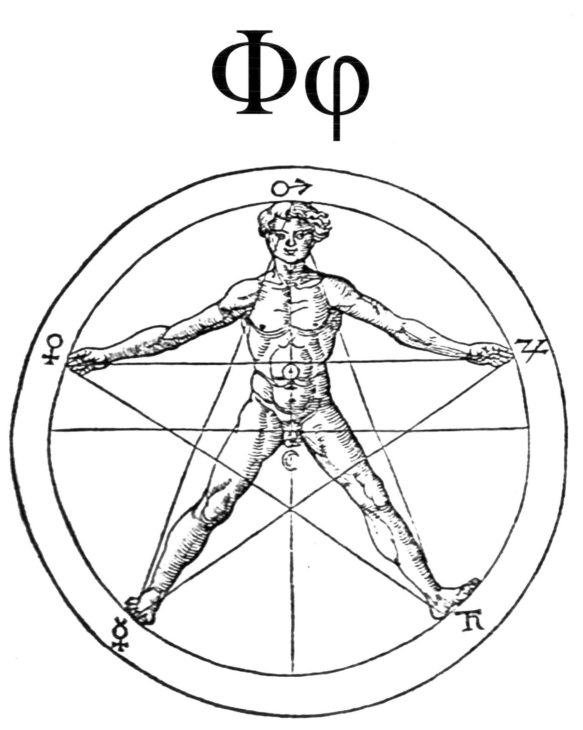

THE MÖBIUS STRIP

August Ferdinand Möbius (1790–1868) was a German inventor, mathematician and astronomer. He is famous for the discovery of the Möbius strip, a two-dimensional surface with a single face and edge, which is a classic example of a non-orientable surface (the right becomes left and vice versa). It was independently discovered by Johann Benedict Listing around the same time.

This figure, known as the Möbius strip or band, can be constructed in three dimensions joining together the two ends of a strip of paper after twisting one of these ends 180°. The Möbius strip has only one face. At first glance the strip seems to have two faces, though the fact that it has one is easily discernible using a pencil—by just drawing a line along the tape, you will see that you get to the starting point without crossing the edge. Another of the intriguing properties of the Möbius strip is that it is not orientable. This means that you cannot distinguish the top from the bottom, or the right from the left.

The infinite Möbius strip has many applications in real life. For instance, think of any band that has to turn held by cylinders in order to transfer rotating movement from one place to another, such as the drive belt in a car or a bicycle chain. With repeat use, the band rubs against the cylinders causing it to wear. If we used a cylindrical band (without the twist), it would only wear on the inside leaving the outside intact. However, with a Möbius band, after one turn, contact would be made with "the other side" (although we know that there is only one side in this case), which would rub on the second turn. This way, both sides wear at the same rate and the band lasts double the time. This system is in use on conveyor belts, recording tapes (so both sides can be recorded on, therefore, double the time), and in many similar applications.

 If the tape is cut in half lengthwise, we do not obtain two tapes of the same size, but one tape twice as long. If we cut the resulting tape again in half lengthwise, we end up with two equal tapes that are linked.

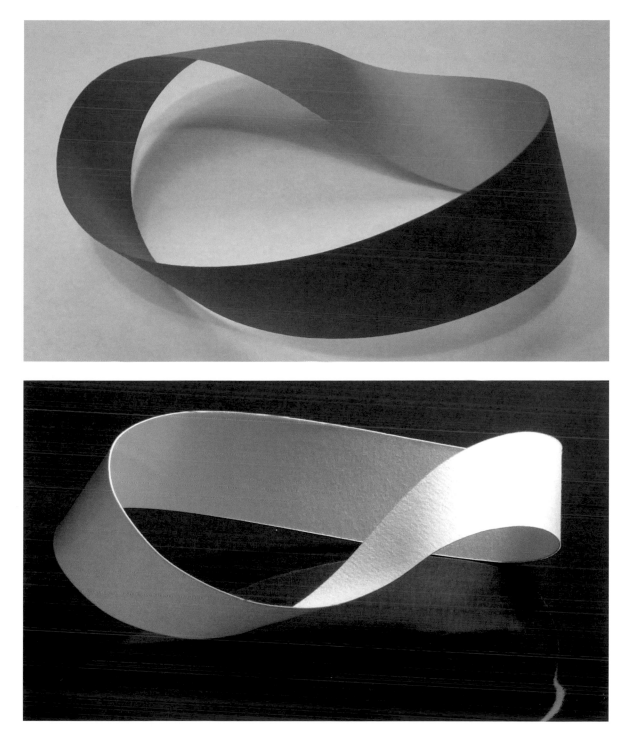

THE ARCHIMEDEAN SPIRAL

The Archimedean spiral, also called uniform or arithmetic spiral, is named after the Sicilian mathematician who lived in the third century BC and who defined it with the following words: "If a straight line one extremity of which remains fixed is made to revolve at a uniform rate in a plane, and, at the same time, a point moves at a uniform rate along the straight line, starting from the fixed extremity, the point will describe a spiral in the plane."

Archimedes described it in his book *On Spirals*. The special features of this spiral are that between two successive turns, the separation distance is the same, and expansion and rotation take place at the same speed (the link between them is linear).

The Archimedean spiral has many real world applications, such as compression springs, formed by two interspersed Archimedes spirals of the same size, which are used to compress liquids and gases. The grooves of the first phonograph recordings form an Archimedean spiral, as they are equally spaced, thus maximizing recording time. Asking a patient to draw an Archimedes spiral is a way of quantifying uncontrollable hand shaking, so it is a basic test to diagnose neurological diseases.

These spirals are also used in projection systems to process digital light in order to minimize the rainbow effect, which makes it appear as if various colors were projected at once when actually quick cycles of red, green and blue are projected.

Finally, it is worth noting that in the book *On Spirals* Archimedes uses the spiral to calculate the length of an arc, to square the circle (constructing a square with the same area of a given circle) and to trisect an angle (dividing it into three equal parts). Although this curve allowed him to address two of the three main concerns, the squaring of the circle and the trisection of the angle, unfortunately for Archimedes, the Greeks demanded the resolution using only ruler and compass, and his curve, the uniform spiral, cannot be constructed using those instruments alone.

Archimedes is one of the leading scientists in classical antiquity. His achievements include advances in hydrostatics, the principle of the lever and the design of several machines.

THE DÜRER SPIRAL

Albrecht Dürer (1471–1528) is the most famous artist of the German Renaissance, known worldwide for his paintings, drawings, engravings and theoretical writings on art that had a profound influence on 16th-century artists in his native country and the Netherlands.

The quality of Dürer's work, the phenomenal amount of art he produced, and the influence he exerted over his contemporaries were of great relevance in the history of art. His interest in geometry and mathematical proportions, his profound sense of history, his observations of nature, and the awareness he had of his own creative potential corresponded to the Renaissance spirit of constant intellectual curiosity.

This great painter and mathematics enthusiast published a book called *Treatise on Mensuration With the Compass and Ruler in Lines, Planes, and Whole Bodies*. It is an interesting book aimed at teaching artists, painters, and mathematicians different ways of drawing geometric figures, among them a complex type of spirals, those based on gnomonic growth—in other words, forms that are obtained by repeatedly fitting similar geometric figures and joining their vertices.

Of these, one would go down in history with his name—Dürer's spiral. This is not an Archimidean spiral or a logarithmic spiral, given that neither of these can be drawn with a ruler and compass; nevertheless it is similar to the latter. It is one of the gnomonic spirals based on the famous golden number, or more precisely, on golden rectangles.

Golden rectangles are made using the golden ratio, whereby the quotient between their longer side and shorter side is the golden number. If a square is added to the longer side of a golden rectangle, another golden rectangle is obtained. If this process is continued and the two opposing angles of each of the series of squares that are added are joined by the arc of a circle, we obtain a Dürer's spiral.

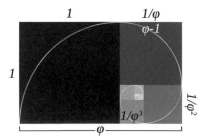

The golden or Dürer spiral is the only one that can be constructed with ruler and compass. It is constructed from a rectangle whose sides form the golden ratio.

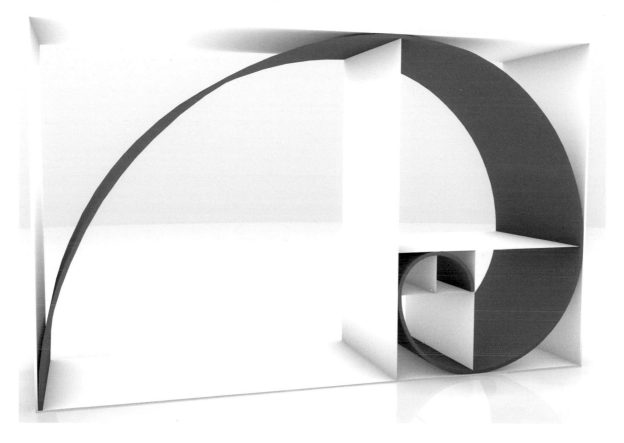

THE LOGARITHMIC SPIRAL

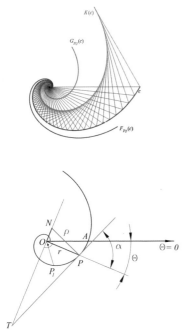

The mathematician Jakob Bernoulli (1654–1705) chose the figure of the logarithmic spiral and the Latin text *eadem mutata resurgo* ("although changed, I shall arise the same") to be used in his epitaph. Contrary to his wish to have a logarithmic spiral (constant in the radius) carved on his tombstone, what the master masons carved on it was an Archimedean spiral (constant in the difference).

The logarithmic spiral is distinguished from the Archimedean one owing to the fact that the distances between turns increase in geometrical progression, while in an Archimedean spiral these distances are constant.

René Descartes (1596-1650) and Bernoulli had begun their study in the 17th century. Descartes found in the logarithmic spiral the solution to a problem posed by Galileo (on the trajectory of a falling body in a rotating earth), a year after the publication of *La Geometrie*, and Evangelista Torricelli (1608–1647), using methods similar to Archimedes, was the first to estimate its length.

These two mathematicians lacked the powerful tool of the calculus of Leibniz and Newton to analyze the logarithmic spiral more thoroughly. This honor went to Jakob Bernoulli in the early 18th century, who devoted an entire book to this spiral, in which he named it *spira mirabilis*, the wonderful spiral, being so excited by its geometric properties.

It's fascinating to see how the logarithmic spiral, also called equiangular or growth spiral, is a kind of spiral curve very common in nature, in the vegetable kingdom, in the forms of galaxies, in the shells of some species of molluscs, in the forms of devastating hurricanes, and even in art since prehistoric times.

Nature offers logarithmic spirals in seashells, the tail of seahorses and the proboscis of butterflies.

THE FIBONACCI SEQUENCE

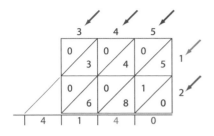

Leonardo of Pisa (c. 1170–1250), also called Fibonacci, has gone down in history as the mathematician who spread across Europe the Hindu-Arabic decimal positional notation (base 10 or decimal) and the use of Arabic numerals including a null digit, zero. Aware of the superiority of Arabic numerals, Fibonacci traveled throughout the Mediterranean countries to study with the leading mathematicians of that time, returning to Pisa around 1200.

In 1202, he published what he had learned in *Liber Abaci*. This included the nine Indian figures and the sign for zero for the first time in the Western world, as well as clear rules to carry out operations with these numbers both with whole numbers and fractions, simple and complex cross-multiplication, rules for calculating the square root of a number, and even instructions for solving first and second degree equations.

However, Fibonacci is best known for an odd infinite sequence of natural numbers:

0, 1, 1, 2, 3, 5, 8, 13, 21, 34, 55, 89, 144...

The sequence, discovered by Indian mathematicians around 1135, was first described in Europe in the *Liber Abaci* by the problem of rabbits. Fibonacci suggested imagining a couple of rabbits, imprisoned in a breeding field and calculating how many pairs of rabbits there will be after a given number of months, knowing that each pair breeds a new pair every month, which in turn would not have rabbits until they reach the fertile age 2 months after birth.

It is easy to see that in this series each term is the sum of the previous two. Each element of this sequence represents the Fibonacci number. There is another curious relationship between the elements: the ratio between each term and the previous one moves ever closer to a very special one, already known to the Greeks and applied in their sculptures and temples—the golden ratio. The Fibonacci sequence has numerous applications in computer science, mathematics and game theory, and appears in biological settings such as tree branches, the arrangement of leaves on stems, the flora of the artichoke and the ratio between the bones of the hands.

In his Liber Abaci, *Fibonacci did away with the use of Roman numerals and fully developed the use of Arabic numerals.*

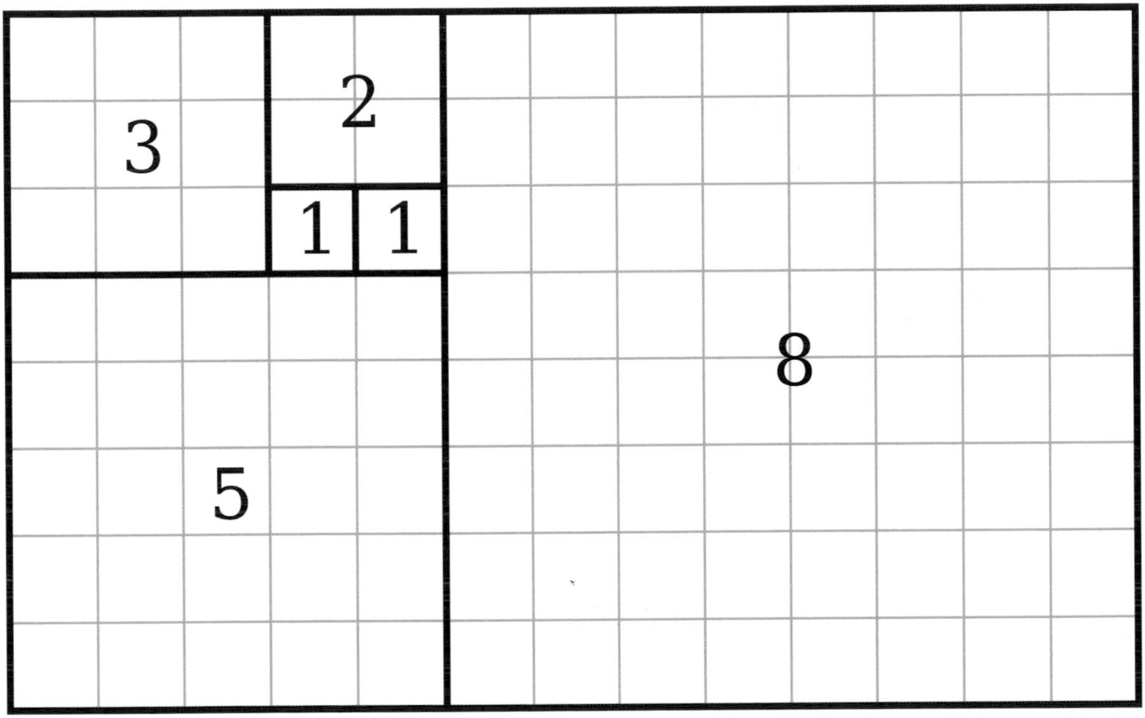
3
2
1
1
5
8

COPY

MANDELBROT FRACTALS

The mathematician Benoit Mandelbrot (1924–2010) was a pioneer of fractal geometry, the mathematics of irregularity. The professor was interested in issues that had never bothered scientists, such as the patterns that are governed by roughness or the cracks and fractures in nature. Mandelbrot argued that fractals, in many respects, are more natural, and therefore better understood intuitively by man than the objects based on classical geometry, whose contours have been smoothed artificially. One of the most popular objects of fractal geometry is the Mandelbrot set.

Mandelbrot coined the term "fractal" from the Latin *fractus*, "broken," in 1975. With a few exceptions, like the eye or the Moon, the forms of nature are rough, irregular and not homogeneous. Up to the study of fractals, mathematicians had always focused on simple figures. For some mathematicians it is too complex to define these forms, as the list of their properties is known but it is difficult to find a universal and absolute description of "fractal." A fundamental property of fractals is self-similarity, which refers to a certain invariance with respect to scale, or put another way, by examining certain parts of a structure or image the structure or entire image reappears in miniature. The same motif appears at different scales, at an infinite number of scales.

The relationship of fractals with infinity is peculiar. It illustrates the paradox stated by Mandelbrot in the article "How long is the coast of Britain?" published in the journal *Science* in 1967, which states that anyone who tries to measure the coast will get a different outcome depending on the scale, the degree of detail. The study does not conclude that any coastline or geographic edge is really fractal, which would be physically impossible—the conclusion was simply that the distance measured from one coast can empirically behave like a fractal in a set of measurement scales. In an ideal fractal, the coast—or actually any rough outline—would be infinite.

To express the presence of fractals in nature, Mandelbrot used the following example: "Clouds are not spheres, mountains are not cones, coastlines are not circles, and bark is not smooth, nor does lightning travel in a straight line."

PARADOX OF THE GRAND HOTEL

The Grand Hotel is an abstract construction that is involved in several thought experiments invented by German mathematician David Hilbert (1862–1943) that explain, simply and intuitively, paradoxical facts related to the mathematical concept of infinity. Hilbert envisioned a hotel with infinite rooms numbered 1, 2, 3, 4 ... to infinity. Remember that "infinite" does not mean "a large number." If so, you could always find a larger number, a large number +1. That said, we welcome you to the Grand Hotel.

As soon as they opened the doors of this hotel, people began to crowd it and the hotel with infinite rooms was full of infinite guests. This led to the first paradox, so a rule was established that guests would always have a room secured but with the prior agreement that they would have to change rooms every time they were asked.

A man came to the hotel when it was full. The customer was not worried about this, because the Grand Hotel claimed that everyone would have a room. He requested his room and the receptionist, knowing that there would be no problem, advised all guests to add one to their room number and move to that room. Thus, the new guest was able to stay in room number 1. And the guest in the last room? Simply put, there is no last room

With the hotel full again, a travel rep arrived: his problem was that he had an infinitely large bus full of infinite tourists who needed to stay overnight at the hotel. It was thus a matter of making room for infinite guests in a hotel with infinite rooms, all occupied at the same time. But the receptionist had no problem. He took to the microphone and asked all guests to move into the room number corresponding to the result of multiplying your current room by 2. Thus, all guests moved to a room with an even number, and all uneven rooms were free. As there exists an infinity of odd numbers, the infinite tourists could stay without problems.

With the hotel full, another representative from the agency arrived even more worried than the first: now the agency had an infinite number of buses full of an infinite number of tourists on each one. The receptionist remained impassive as he took to the microphone and communicated with the rooms whose number was prime or a power thereof; he asked them to raise number 2 to the room number (n) which they currently occupied (2^n) and they were to change to that resulting room number. Then he assigned a prime number (greater than 2) to each bus full of people and an odd number to every tourist within each coach, so that the room for each of the tourists was calculated by taking the prime of the bus number (p) and raising it to the number of occupants assigned within the bus (t), giving p^t. As there are an infinite number of primes and an infinite number of odd numbers, an infinite number of infinite guests were able to lodge in this astonishing hotel of infinite rooms.

This thought experiment is related to Cantor's transfinite numbers, which serve to measure the magnitude of various infinite sets.

Skip

TORRICELLI'S TRUMPET

$y = 1/x$

$\rightarrow \infty$

Evangelista Torricelli (1608–1647) was an Italian mathematician and physicist. Orphaned at an early age, he was educated under the tutelage of his uncle, Jacopo Torricelli, a monk who taught humanities. In 1627, he was sent to Rome to study science with the Benedictine Benedetto Castelli (1578–1643), professor of mathematics at the University of Rome La Sapienza and one of the first disciples of Galileo.

In addition to discovering the principle of the barometer in 1643, which demonstrated the existence of atmospheric pressure and the fundamental principles of hydromechanics, Toricelli also devised the curious Gabriel's horn or Torricelli's trumpet, a geometric figure that has an infinite surface but a finite volume. This apparent paradox can be illustrated by noting that one would need an infinite amount of paint to cover the interior surface, while it would be possible to cover the whole figure with a finite amount of paint and thus cover the surface.

Mathematically, it explains why the surface generated by the curve $y = 1/x$ (hyperbola), turning on the x-axis (with x between 1 and $+\infty$) is infinite, and yet the volume defined by the surface is finite. To reach this conclusion, the volume is calculated as the sum of infinite areas of circles with radius y (as it rotates around the x-axis) and taking into account that $y = 1/x$, which results in a finite volume. For the surface, all one need do is sum the perimeters of the circles with radius y, which is an infinite surface.

Thus, the solution of the paradox is that an infinite area requires an infinite amount of paint if the layer has a constant thickness. This is not applicable for the interior of the horn, since most of the length of the figure is not accessible for painting, especially when its diameter is smaller than a molecule of paint. If we consider a layer of paint of normal thickness, an infinite amount of time would be necessary before reaching the "end" of the horn.

In other words, there would come a time when the thickness of the trumpet would be smaller than a molecule of paint with which, let's say, a drop of paint would cover the rest of the surface of the trumpet (even if it were infinite). Thus, the fact that the surface of the trumpet is infinite does not imply that the amount of paint needed is infinite.

Torricelli's trumpet is a geometric figure that has the characteristic of having an infinite surface and a finite volume.

SLIP

PEDRO NUNES AND THE LOXODROME

A rhumb line or loxodrome (*loxos* = oblique, *dromo* = path) is a curve joining two points on the Earth's surface by cutting all meridians at the same angle. In navigation, it is widely used because it corresponds to the course marked by a constant position of the compass. Its representation on a map will depend on the type of projection, for example, on Mercator's it is a straight line, but on the globe it is a spiral.

This curve was studied by the Portuguese geographer Pedro Nunes (1502–1578), who published his discovery in 1546 in *Navigation Treaty*. Previously it was believed that, to traverse the Earth's surface with a fixed rhumb, that is, to form a constant angle with the meridian, the covered line was considered a maximum circle or, in other words, a ship that theoretically followed this path was imagined to travel the full way around the world, returning to the starting point.

Nunes was the first to point out the fallacy in this deeply ingrained concept, rigorously proving that, far from ever reaching the starting point, the curve gets closer to the pole, going endlessly around it without ever reaching it, or, in technical language, the curve and the pole are at asymptotic points, because the distance between the two points on this curve that are on the same meridian decreases as the latitude increases from the Equator to the pole.

Pedro Nunes, one of the most important mathematicians, astronomers and geographers of the 16th century, discovered the shortest twilight and invented the nonius, an instrument for measuring lengths that allows measuring angle degree fractions using an auxiliary scale.

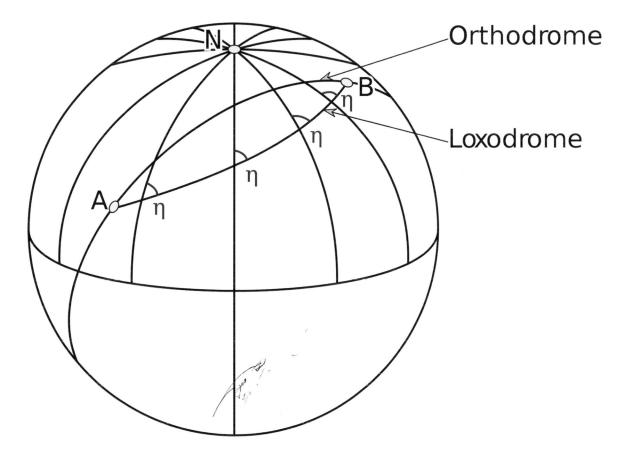

MATHEMATICS   **AVOGADRO'S NUMBER**

$$N_{\mathrm{A}} = 6.022\,141\,29\,(27) \cdot 10^{23}\ \mathrm{mol}^{-1}$$

Amedeo Avogadro (1776–1856), Earl of Quaregna and Cerreto, was an Italian physicist and chemist, professor of Physics at the University of Turin from 1834 to 1850. He formulated the eponymous Avogadro's law, which states that equal volumes of different gases under the same conditions of temperature and pressure contain the same number of particles. He found that simple gases such as hydrogen and oxygen are diatomic (H_2, O_2), and he assigned the H_2O formula for water.

The number of particles in one "mole" of substance was named Avogadro's contant or number (N_A) in his honor. Its value is 6.023×10^{23}, a very large number close to a quadrillion: 602,300,000,000,000,000,000,000.

The definition of "mole" clarifies the meaning of the physical quantity known as "quantity of matter," which may be distinguished from mass: matter consists of elementary entities (atoms, ions, molecules, electrons, etc.) in the sense that in a particular process, these are the elementary entities involved as subjects. Two samples of matter, formed, for example, by atoms, ions or molecules have the same amount of matter if they contain the same number of atoms, ions or molecules, respectively. The mole establishes the number of these entities that are considered a unit: a mole contains N_A elementary entities. Thus, a mole of hydrogen atoms contains 6.023×10^{23} atoms of hydrogen and one mole of hydrogen molecules contains 6.023×10^{23} molecules of hydrogen. When the mole is used as the unit of quantity of matter, the considered elementary entities must be specified (one mole of atoms, one mole of molecules, etc.).

The accuracy of the definition of the mole makes concepts like atom-gram (molar mass of atoms) and mole-gram (mass of one mole of molecules) unnecessary, which are calculated from the atomic weight and molecular weight, respectively, of a given substance.

Avogadro's number indicates how many atoms of an element there are in a certain amount whose weight in grams coincides with the atomic mass of the element. The atomic mass of carbon is approximately 12, so 12 grams of carbon will contain N_A carbon atoms.

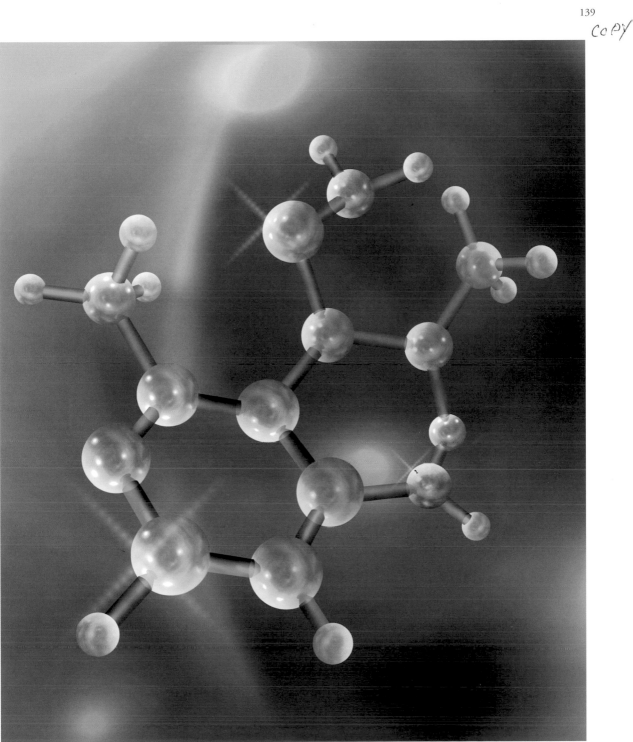

THE GOSPER CURVE

Bill Gosper (1943) is an American mathematician and programmer. Along with Richard Greenblatt, he is considered the founder of the hacker community and occupies a privileged place within the Lisp community. He is also known for his work on continued fraction representations of real numbers and for suggesting the algorithm to find closed forms of hypergeometric identities (which now bear his name).

In 1973, he studied a variant of the Koch island whose generator is a hexagon. Its main feature is that of tiling, that is, it can cover the whole plane, and also, like all fractal objects, it is self-similar. For this island, Gosper built the elegant space-filling curve, which is named after him. The Gosper curve is obtained as a limiting curve of a sequence of broken lines, from a segment that, with each iteration, folds in different directions. The space determined by the Gosper curve is a fractal called a Gosper island that looks like a gear or a snowflake.

Each island is divided into seven regions (hexagons) related to the entire ensemble for a similar reason. In fact, you can link together seven copies of the Gosper Island to form a similar shape but √7 times larger than in the two dimensions of the plane. Although you cannot build a larger hexagon juxtaposition of several hexagons, the Gosper island deforms enough to allow this subdivision into 7 parts. Similarly, a Gosper island can be extended to an infinite curve that covers the plane.

Although the surface of the seven islands that cover the plane is seven times greater than that of the central island, its perimeter is 3 times greater than the central island, but if calculated using the equations of traditional scale, it should be only about 2.6 times greater.

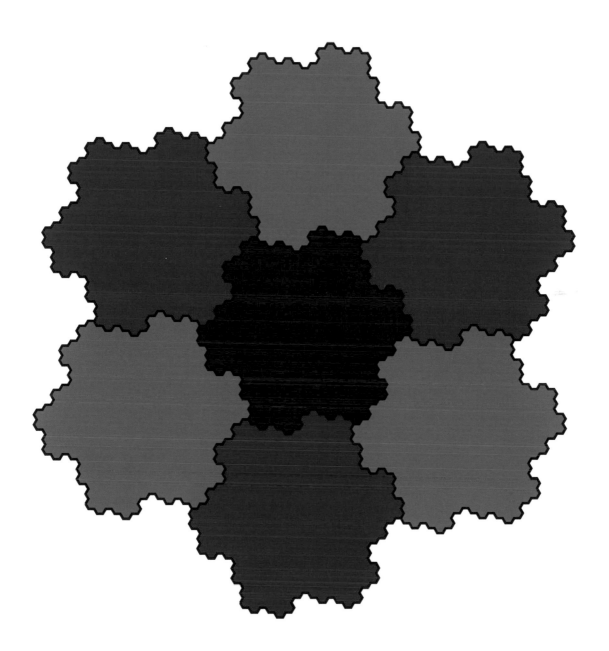

MATHEMATICS **THE LEMNISCATE**

The English mathematician John Wallis (1616–1703), whose works form a precursor of calculus, introduced the use of the symbol ∞ to represent the notion of infinity in his work *Arithmetica infinitorum* (1655). This symbol is known as the lemniscate.

In mathematics, a lemniscate is a type of curve described by the following equation in Cartesian coordinates:

$$(x^2 + y^2)^2 = a^2(x^2 - y^2) = x^4 + 2x^2y^2 + y^4$$

The graphical representation of this equation generates a curve similar to ∞.

The lemniscate was first described in 1694 by Jakob Bernoulli (1654–1705) as the modification of an ellipse, a curve that is defined as the locus of points such that the sum of the distances from two focal points is a constant. In contrast, a lemniscate is the locus of points such that the product of these distances is constant. Bernoulli called it *lemniscus*, which in Latin means "pendant ribbon."

The lemniscate is the inverse transformation of a hyperbola, with the inversion circle centered at the center of the hyperbola.

The determination of the arch length of the lemniscate led to the discovery of elliptic integrals in the 18th century. Around 1800, the elliptic functions involved in these integrals were studied by Carl Friedrich Gauss. They would not be published until much later, but he referred to them in the notes of his work *Disquisitiones Arithmeticae*. The basis of the reticle defined by the fundamental pair of periods (ordered pairs of complex numbers) has a very special form, being proportional to the Gaussian integers. For this reason, the set of elliptic functions with the complex product by the imaginary unit is called a lemniscate set.

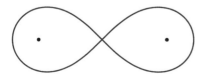

$$a^2(x^2 - y^2)$$
$$a^2x^2 - a^2y^2$$

$$(x^2 + y^2)$$
$$\cdot (x^2 + y^2)$$

$$= x^4 + x^2y^2$$
$$+ x^2y^2 + y^4$$
$$\overline{x^4 + 2x^2y^2 + y^4}$$

Throughout his life, John Wallis made significant contributions to trigonometry, calculus, geometry and analysis of infinite series.

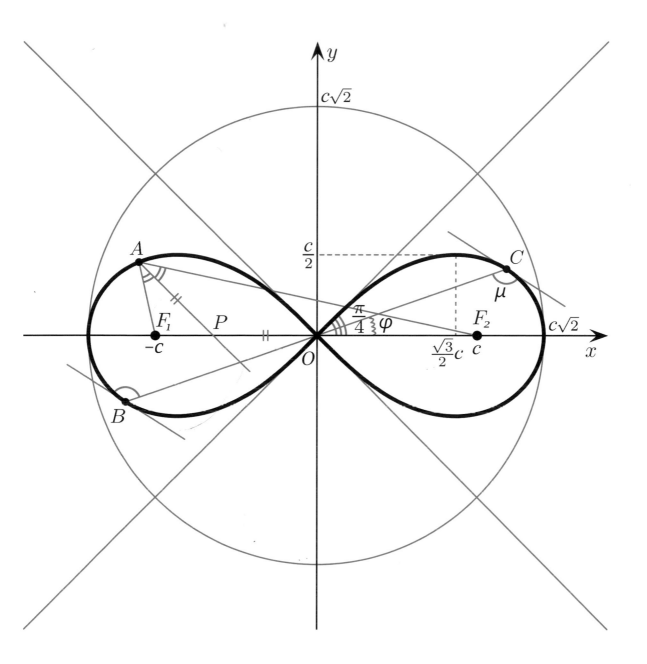

SKIP

DIVISION BY ZERO

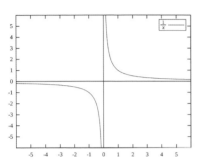

In mathematics, a division by zero is a division in which the divisor is zero; yet zero is the only real number by which we cannot divide. The reason is that 0 is the only real number that has no multiplicative inverse, since the product $n \cdot 0 = 0$.

The problem arose in mid-17th century, when the use of zero and negative numbers began to become popular in India. The first to try and solve this problem was the Indian mathematician Bhaskara I, who deduced that $\frac{n}{0} = \infty$.

The expression $\frac{n}{0}$ is an indefinition. However, if $n = 0$, we obtain the expression $\frac{0}{0}$, which is an indetermination, meaning that the operation of dividing a number by 0 is known as indefinition, if the dividend is different from 0 (that is, no return value would be valid), or indetermination, if the dividend is 0 (any value returned would be valid). Expressions such as 8:0 and 0:0 are thus meaningless, because it is nonsensical to "distribute" 8 between no number or share out nothing between nothing. This is an intuitive idea and common sense is enough to answer this problem.

From the viewpoint of mathematical analysis, the indefinition of a division by zero can be solved through the concept of limit. Suppose we have the following expression:

$$f(x) = \frac{n}{x}$$

where n is a natural number (non-zero and infinite). So to calculate the value of $f(0)$, you can use an approximation of the limit from the right:

$$f(0) \simeq \lim_{x \to 0^-} \frac{n}{x} = +\infty$$

or from the left:

$$f(0) \simeq \lim_{x \to 0^-} \frac{n}{x} = -\infty$$

When the value of x tends to zero, n/x reaches an immensely large value (positive or negative). It is usually expressed by saying that when x tends to zero, n/x moves closer to infinity:

$$f(0) = \frac{n}{0} \simeq \infty$$

However, although it is apparently acceptable in theory, this solution can generate mathematical paradoxes, known as different infinities.

In computing, particularly programming, a division by zero is considered a classic logical error. Since many classical division computer algorithms use the successive subtraction method, as the zero divisor, the remainder continues to run indefinitely, since the dividend never changes. The application then runs an infinite loop.

THE THEORY OF PROBABILITY

$$P(A_1 \dot\cup A_2 \dot\cup \cdots) = \sum P(A_i)$$

$$0 \leq P(A) \leq 1$$

The theory of probability is the branch of mathematics that studies the possibilities of random events occurring, in order to predict possible outcomes, combining scientific certainty with the uncertainty of chance.

The theory of probability has evolved through different concepts of probability. On the one hand, classical probability theory is based on considering that any of the possible outcomes are equally likely to occur (equiprobable events). This understanding of probability is thus *a priori*. When a random experiment has n possible outcomes, all equiprobable, we use the classical method to estimate the probability of the occurrence of each one. Therefore, each result will have an equal probability of $1/n$. Imagine for example a 6-sided die. The probability of rolling the number 5 is equal to the probability of any other value: $1/6$.

On the other hand, frequentist theory defines probability as the observed relative frequency of an event for a large number of attempts, or as the fraction of times an event occurs in the long run, when conditions are stable. This method uses the relative frequency of past occurrences of an event as a probability. We determine how often something has happened in the past and use that number to predict the probability of it happening again.

If a random experiment is repeated several times and under the same conditions, the ratio between the number of times to produce a result and the total number of times the experiment is carried out tends toward a fixed number. This property, established by Jakob Bernoulli, is known as the law of large numbers. An estimate value is calculated with the method of relative frequencies, rather than an exact value. As the number of observations increases, the estimates tend to approach the real probability. For example, the more times you throw a coin, the more probable it is that you will achieve 50 percent heads and 50 percent tails. If you could throw it infinite times, you would end up with an exact result of 50–50.

Probability theory began with the analysis of games of chance, but currently has applications in fields as diverse as economics and finance, quantum physics and statistics.

use?

LEGO, INFINITE INTERCONNECTABILITY

LEGO is a Danish toy company famous worldwide for its interlocking plastic bricks. The name LEGO was adopted by the company in 1932, and derives from the Danish phrase *leg godt*, meaning "play well."

On January 28, 1918, Ole Kirk Christiansen opened a carpenter's workshop in Billund, and earned his living building houses and furniture for farmers in the area with the help of a small team of apprentices. His workshop burned down in 1924 and Ole Kirk saw the disaster as an opportunity to build a larger shop that would allow him to expand his business. Looking for ways to minimize production costs, Ole Kirk began producing scale models of their products as a design model. Their miniature ladders and ironing boards inspired him to produce toys.

From 1932 to 1947, LEGO was devoted almost exclusively to producing wooden toys. Then, with the popularization of plastic, it began to manufacture the legendary interlocking bricks that led the company to worldwide fame. Made from cellulose acetate, they were developed in the style of traditional wooden blocks, which could be stacked on each other, but with the benefit of secure connections through the innovative design of their interlocking nodes and hollows.

In 1969 he introduced the DUPLO system, designed for young children. DUPLO bricks are much larger than the LEGO bricks and therefore safer. Yet both systems are compatible: LEGO bricks can be connected easily with DUPLO bricks, making the transition to LEGO much simpler when children get older.

One of the basic features of LEGO bricks throughout their history has been that each one is, above all, part of a system. Each new configuration and series is fully compatible with the rest of the system, the LEGO pieces fit with all others in some orientation, regardless of size, form or function. To give us an idea of their interconnectivity, with only 8 pieces of 2 × 4 there are some 8,274,075,616,387 combinations possible.

There are 2,000 different pieces in 55 colors. The most common are red, yellow, blue, white and light gray. LEGO did not manufacture green bricks for a lengthy period, to prevent them being used to build military vehicles and their blocks being converted into a war toy.

THE GOOGOL AND GOOGOLPLEX

$Googol_{(n)} = n^{(n^n)}$

$Googol_{(2)} = 2^{(2^2)} = 2^4 = 1.00.00_{(2)} = 16_{(10)}$

$Googol_{(3)} = 3^{(3^3)} = 3^9 = 1.000.000.000_{(3)} = 19.683_{(10)}$

$Googol_{(4)} = 4^{(4^4)} = 4^{16} = 1.0000.0000.0000.0000_{(4)} = 4.294.967.296_{(10)}$

$Googol_{(10)} = 10^{(10^{10})} = 10^{100}$

A googol is a cardinal number that is 1 followed by 100 zeros, or in scientific notation, 10^{100}. It is roughly equivalent to the factorial of 70, and its only prime factors are 2 and 5 (100 times each). In the binary system it would occupy 333 bits.

The googol is of no particular importance in mathematics and has no practical uses. The term googol was coined in 1938 by Milton Sirotta, a 9-year-old nephew of American mathematician Edward Kasner, who introduced the concept in his book *Mathematics and the Imagination*. Kasner created it to illustrate the difference between an unimaginably large number and infinity, and sometimes it is used this way in the teaching of mathematics.

Kasner went a step further and defined the googolplex as the number 10 to a googol (10^{googol}), a huge number, truly difficult to fathom, which is a 1 followed by a googol of zeros. From a physical standpoint, a googol is greater than the number of atoms in the known universe (which only amounts to 10 to the power of 78), so even if we wanted we could not write the figure of googolplex because a sheet of paper large enough to write out all the zeros of this number does not fit inside the universe.

Indeed, the resemblance of the name googol with the name of the famous search engine is no accident. The original founders of Google were actually going to call it Googol, a name that was chosen for its intention to cover a huge number of websites, but ended up with Google due to a misspelling by Larry Page, though some recollect that it was a graduate student, Sean Anderson, who made the error. It is fitting, then, that Google's headquarters is called Googleplex.

1 googol = 10^{100} = 10,000,000,000,000,000,000,000,000,000,000,000,000,000,
000,000,000,000,000,000,000,000,000,000,000,000,000,000,000,000,000,000,000

10 x 33 sets of 3 zeros

Edward Kasner (1878–1955) became professor emeritus of the Department of Mathematics at Columbia University. He devised the googol to show how huge infinity is through a number so large that it is unimaginable but it is still not even close to infinity.

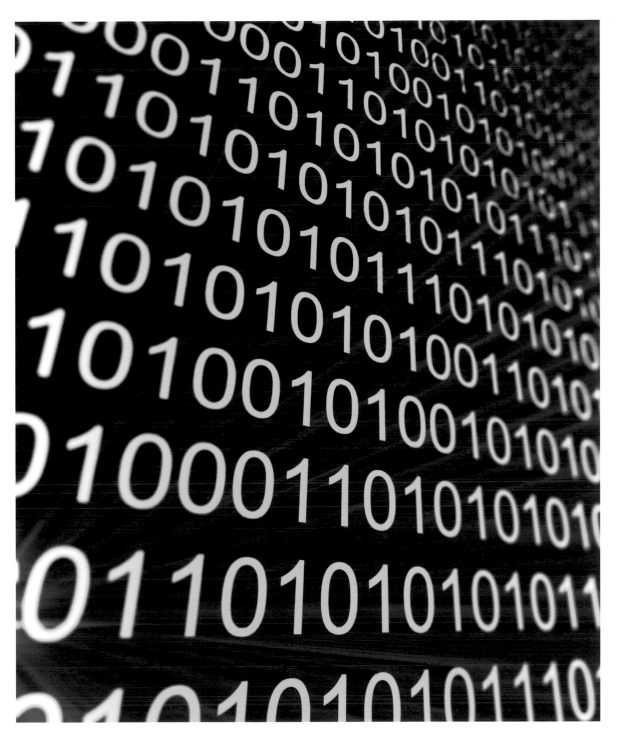

EUCLID'S THEOREM

Euclid was a Greek mathematician (c. 325 BC–265 BC) and is known as the father of geometry. Few details of his life are known: he probably studied in Athens and lived in Alexandria, where he founded a mathematical school. Without a doubt, Euclid's best-known work is *Elements*. It is said that after the Bible, it is the most widely published book throughout history and has been, until recently, a bedside book in the teaching of elementary geometry in almost all schools in the world.

Elements consists of 13 books. The first six deal with elementary geometry. The next four are devoted to matters of arithmetic, and include well-known results such as the algorithm that calculates the greatest common divisor of two numbers, and of the most famous demonstrations of the Pythagorean Theorem. The last three volumes are devoted to solid geometry. Euclid used a logical structure for the demonstrations, thus laying the foundations of which, thereafter, would be the classic way to establish a mathematical proposition. It was the first treatise on mathematics in recorded history.

Euclid was the first to prove that a set consisting of prime numbers is infinite, in proposition 20 of Book IX of *Elements*. He did it by the method of *reductio ad absurdum*, assuming there is a prime number p which is the last prime number. Euclid's proof of the theorem is simple: assume that p is the largest prime number and we build another number q obtained by multiplying all the prime numbers to the last, p, and then we add 1.

$$q = (2 \cdot 3 \cdot 5 \cdot 7 \cdot \cdots \cdot p) + 1$$

Clearly q is not divisible by any prime, as it would yield a remainder of 1, then q is divisible only by 1 and itself, that is, q would be prime. However, q is greater than p, and so p is not the largest prime number. Therefore, there cannot exist a prime number that is the greatest and thereby we verify the existence of infinite prime numbers.

Since the existence of infinite prime numbers was demonstrated, mathematicians of all eras have tried to find a formula to generate prime numbers, but have thus far failed.

7, maybe

THE MERSENNE PRIMES

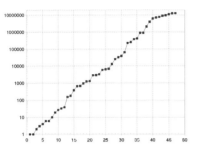

Marin Mersenne (1588–1648) was a monk who was intensely devoted to the study of science, following the footsteps of Galileo, whose work was published in the books *Les Méchaniques of Galileo* (1634) and *Nouvelles Pensées de Galilée* (1630). Founder of the Parisiensis Academy, his school formed a communication hub among European scholars of his time (Descartes, Fermat, Pascal, etc.) and that communication helped to develop the mathematics of his time. Although his main occupation was not math (he was more interested in theology, philosophy and music), he left us with several contributions relating to science and he is known for original contributions in the fields of mechanics (laws of the pendulum) and acoustics.

He was the author of *Cogitata Physico-Mathematica*, a work that introduces the Mersenne numbers and Mersenne primes. It says that a number M is a Mersenne number if it is one less than a power of 2 ($M_n = 2^n - 1$). A Mersenne prime is a Mersenne number which is prime. For example, 7 is a Mersenne prime as it fulfills the following ($7 + 1 = 8 = 2^3$). He also compiled a list of Mersenne primes with exponents less than or equal to 257, and surmised that they were the only ones. His list only proved partially successful, errors included M_{67} and M_{257}, which are compounds, and omitted M_{61}, M_{89}, and M_{107}, which are primes, and his conjecture would prove false with the discovery of larger Mersenne primes. He provided no indication of how he came up with that list, and the rigorous verification was only completed two centuries later.

Today 47 known Mersenne primes numbers have been established, and the greatest of all is $M_{43,112,609} = 2^{43,112,609} - 1$, a figure of nearly thirteen million numbers. The largest known prime number at a given date has almost always been a Mersenne prime. The twelve largest Mersenne primes have been discovered through the project GIMPS (Great Internet Mersenne Prime Search), which uses volunteers' computers around the world.

$$M_{(n)} = M_{ab} = 2^{ab} - 1 = (2^a)^b - 1^b = (2^a - 1)\sum_{k=0}^{b-1}(2^a)^k 1^{b-1-k} = (2^a - 1)\cdot\left(1 + 2^a + 2^{2a} + 2^{3a} + \dots + 2^{(b-1)a}\right)$$

The nine largest known primes are named after Mersenne. In addition, the formula of the Mersenne numbers is very simple, so relatively simple algorithms exist.

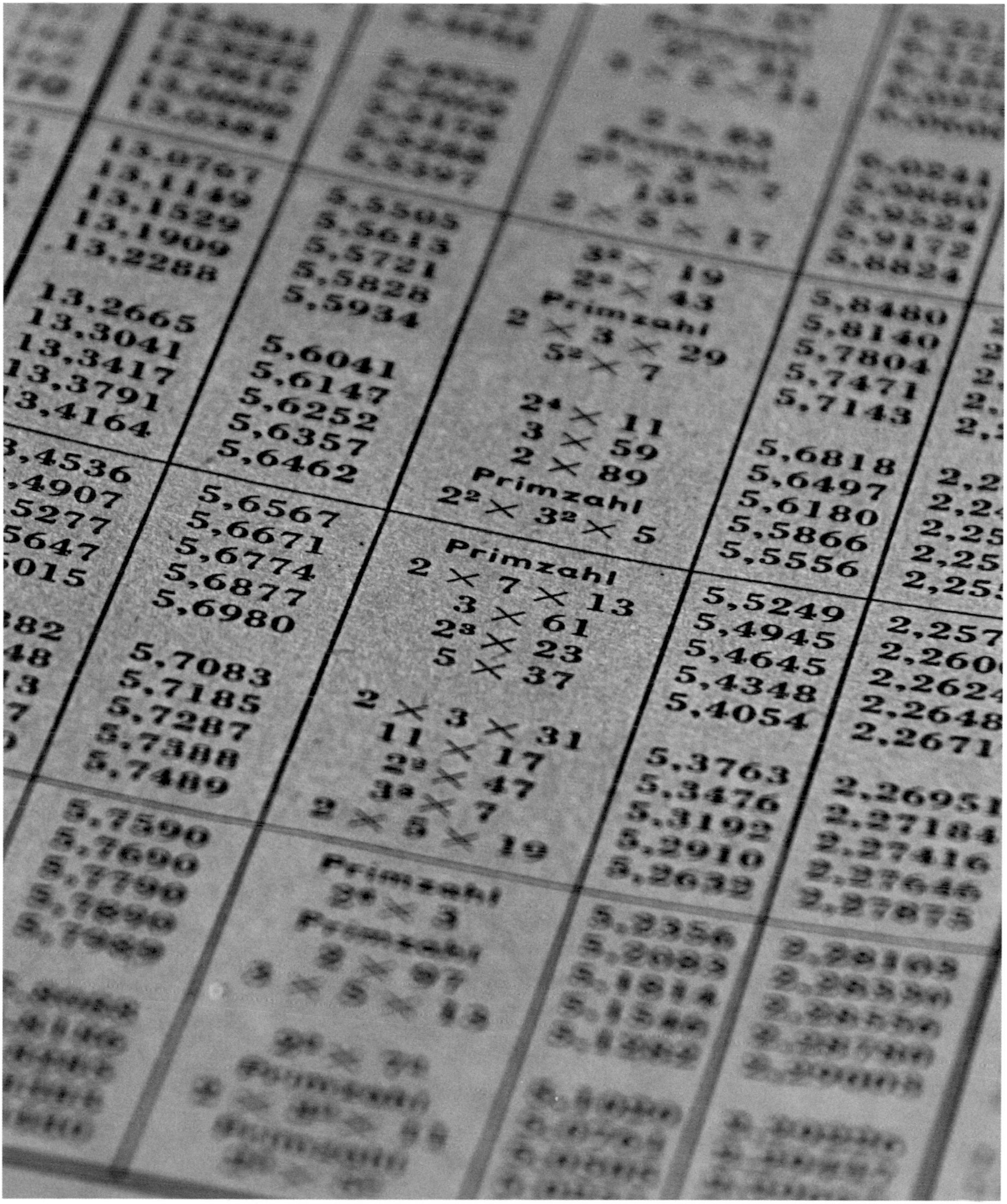

13,0767
13,1149
13,1529
13,1909
13,2288

13,2665
13,3041
13,3417
13,3791
13,4164

3,4536
,4907
5277
5647
015

82
48
13

5,7590
5,7690
5,7790
5,7890
5,7989

5,5505
5,5613
5,5721
5,5828
5,5934

5,6041
5,6147
5,6252
5,6357
5,6462

5,6567
5,6671
5,6774
5,6877
5,6980

5,7083
5,7185
5,7287
5,7388
5,7489

5,5397

2×83
Primzahl
$2^? \times 3 \times 7$
$2 \times 5 \times 17$

$3^2 \times 19$
$2^? \times 43$
Primzahl
$2 \times 3 \times 29$
$5^2 \times 7$

$2^4 \times 11$
3×59
2×89
Primzahl
$2^2 \times 3^2 \times 5$

Primzahl
$2 \times 7 \times 13$
3×61
$2^3 \times 23$
5×37

$2 \times 3 \times 31$
11×31
$2^? \times 17$
$3^? \times 47$
$2 \times 5 \times 19$

Primzahl
$3^? \times 3$
Primzahl
2×97
$3 \times 5 \times 13$

5,8480
5,8140
5,7804
5,7471
5,7143

5,6818
5,6497
5,6180
5,5866
5,5556

5,5249
5,4945
5,4645
5,4348
5,4054

5,3763
5,3476
5,3192
5,2910
5,2632

5,2356
5,2083
5,1814
5,1546
5,1282

6,0241

2,2
2,2
2,24
2,25
2,25
2,25

2,257
2,260
2,2624
2,2648
2,2671

2,26951
2,27184
2,27416
2,27646
2,27875

SQUARING THE CIRCLE

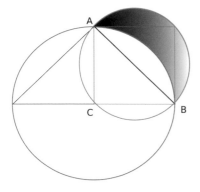

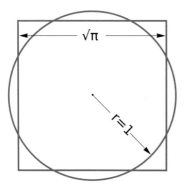

Squaring the circle is a problem that fascinated mathematicians of classical Greece, and which continued to fascinate scholars for 2,000 years. The solution was obtained in 1882 by the German mathematician Ferdinand Lindemann (1852–1939), who was able to show that the number π has infinite decimal places.

Let's take a look at the relationship between the nature of the number π and the quadrature of the circle. It turns out that squaring the circle suggests the possibility of creating a square with the same perimeter as a circle, that is, the concept entails that we should be able to surround a circumference with a string, measure the length of the string and divide it into 4 to be able to form the four sides of a square with these measurements, and that the square would be contained at a point between the inscribed square and the circumscribed square. However, though the concept seems so simple and though it should have a logical resolution, in practice it has proven to be something completely impossible.

This is because the circumference of a circle is $2\pi r$, and π is a transcendental number, irrational and with infinite decimals, which is not the result of any root. Thus, it can never be expressed as the combination of finite numbers, fractions, roots, or figures, nor could it be used to build using geometric tools such as a ruler and compass. With the demonstration of the irrationality of π it is clear that no one should strive to build this square because it is an impossible task. The suspicion of its insolubility has existed since the time of Archimedes but now its impossibility has been absolutely proven.

The use of the Greek letter π for this constant comes from the initial of the Greek words περιφέρεια *(periphery) and* περίμετρον *(perimeter).*

use

THE UNPUBLISHED THEOREM BY PIERRE DE FERMAT

Pierre de Fermat was a 17th-century lawyer and numbers enthusiast who made many valuable contributions to mathematics. His work covered many aspects of mathematics ranging from the reconstruction of some demonstrations by Greek mathematician Apollonius to some innovative contributions to the algebraic method.

There is one detail of his work that never fails to surprise; Fermat used to write the solutions to problems that he developed in the margins of his books. One infamous note was found in his copy of the Greek text *Diophantus Arithmetica*: "It is impossible to separate a cube into two cubes, or a fourth power into two fourth powers, or in general, any power higher than the second, into two like powers. I have discovered a truly marvelous proof of this, which this margin is too narrow to contain."

This notation became one of the most important statements to be proven in mathematics. The problem left mathematicians on edge for over three centuries, it is said that Euler asked for Ferma's house to be raided in search of the solution, but it was never found.

$$x^2 + y^2 = z^2$$

E.g. $3^2 + 4^2 = 5^2$

It is known that Fermat came to prove the case $n = 4$ by the method of infinite descent, and it is likely that he thought that this solution would suffice for general cases, or that he had given up on finding a solution and, as his margin notes were of personal use, there was no need to correct them.

In 1995, Britain's Andrew Wiles finally established the proof to what is known as Fermat's last theorem, using mathematical tools that emerged long after Fermat's death; this means Fermat must have found the solution in another way.

In mathematics and proof theory, the infinite descent is a method to prove a statement about natural numbers, which consists in saying that none of the natural numbers of a certain subset satisfies a certain property.

interuallum numerorum 2. minor autem
1 N. atque ideo maior 1 N. + 2. Oportet
itaque 4 N. + 2. triplos esse ad 2. & ad-
huc superaddere 10. Ter igitur 2. adfci-
tis vnitatibus 10. æquatur 4 N. + 4. &
fit 1 N. 3. Erit ergo minor 3. maior 5. &
farisfaciunt quæftioni.

ζ' ἰνὸς ὁ ἄρα μείζων ἴσως ζ' ἰνὸς μ² β̅. δεή-
σει ἄρα ἀριθμὸς δ' μονάδας δ' τριπλασίονας
ᾗ μ̅ β̅. ἔτι ὑπερέχειν μ̅ ι̅. τρὶς ἄρα
μονάδες β̅ μ̅ ι̅. ἴσαι εἰσὶν ς̅ δ̅ μονάσι
δ̅. κỳ γίνεται ὁ ἀριθμὸς μ² γ̅. ἔτι ὁ μὲν ἐλάσ-
σων μ² γ̅. ὁ δὲ μείζων μ̅ ι̅. κỳ ποιοῦσι τὴ
πρόβλημα.

IN QVAESTIONEM VII.

CONDITIONIS appofitæ eadem ratio eft quæ & appofitæ præcedenti quæftioni, nil enim aliud requirit quàm vt quadratus interualli numerorum fit minor interuallo quadratorum, & Canones iidem hic etiam locum habebunt, vt manifeftum eft.

QVÆSTIO VIII.

PROPOSITVM quadratum diuidere
in duos quadratos. Imperatum fit vt
16. diuidatur in duos quadratos. Ponatur
primus 1 Q. Oportet igitur 16 − 1 Q æquales effe quadrato. Fingo quadratum à nu-
meris quotquot libuerit, cum defectu tot
vnitatum quod continet latus ipfius 16.
efto à 2 N. − 4. ipfe igitur quadratus erit
4 Q. + 16. − 16 N. hæc æquabuntur vnitatibus 16 − 1 Q. Communis adiiciatur
vtrimque defectus, & à fimilibus aufera-
tur fimilia , fient 5 Q. æquales 16 N. & fit
1 N. ⁱⁱ Erit igitur alter quadratorum ⁱⁱⁱ.
alter verò ⁱⁱⁱ & vtriufque fumma eft ⁱⁱⁱ feu
16. & vterque quadratus eft.

ΤΟΝ ἐπιταχθέντα τετράγωνον διελεῖν εἰς
δύο τετραγώνους. ἐπιτετάχθω δὴ τ̅ ιϛ̅
διελεῖν εἰς δύο τετραγώνους. κỳ τετάχθω ὁ
πρῶτος δυνάμεως μιᾶς. δεήσει ἄρα μονά-
δας ιϛ̅ λείψει δυνάμεως μιᾶς ἴσας ᾗ τε-
τραγώνῳ. πλάσσω τ̅ τετράγωνον ἀπὸ ὁσων-
δὴ ποτε ἀριθμῶν μ² ὅσων ἐστὶν ἡ τ̅ ιϛ̅
μ̅ πλευρᾶ. ἴσω ςˢ β̅ λείψει μ̅ δ̅. αὐτὸς
ἄρα ὁ τετράγωνος ἴσται δυνάμεων δ̅ μ̅ ιϛ̅
λείψει ςˢ ιϛ̅. ταῦτα ἴσα μονάσι ιϛ̅ λείψει
δυνάμεως μιᾶς. κοινὴ προσκείσθω ἡ λεῖψις,
κỳ ἀπὸ ὁμοίων ὅμοια. δυνάμεις ἄρα ε̅ ἴσαι
ἀριθμοῖς ιϛ̅. κỳ γίνεται ὁ ἀριθμὸς ιϛ̅. πέμπ-
των. ἴσται ὁ μὲν σϛ̅ εἰκοστόπεμπτ. ὁ δὲ μϛ̅
εἰκοστόπεμπτων, κỳ οἱ δύο συντεθέντες ποιοῦ-
σι εἰκοστόπεμπτα, ἤτοι μονάδας ιϛ̅. καὶ ἔσιν ἑκάτερος τετράγωνος.

OBSERVATIO DOMINI PETRI DE FERMAT.

CVbum autem in duos cubes, aut quadratoquadratum in duos quadratoquadrates & generaliter nullam in infinitum vltra quadratum poteftatem in duos eiufdem nominis fas eft diuidere cuius rei demonftrationem mirabilem fane detexi. Hanc marginis exiguitas non caperet.

QVÆSTIO IX.

RVRSVS oporteat quadratum 16
diuidere in duos quadratos. Ponatur rurfus primi latus 1 N. alterius verò
quotcunque numerorum cum defectu tot
vnitatum, quot conftat latus diuidendi.
Efto itaque 2 N. − 4. erunt quadrati, hic
quidem 1 Q. ille verò 4 Q. + 16. − 16 N.
Cæterum volo vtrumque fimul æquari
vnitatibus 16 Igitur 5 Q. + 16 − 16 N
æquatur vnitatibus 16. & fit 1 N. ⁱⁱ erit

ΕΣΤΩ δὴ πάλιν τὸν ιϛ̅ τετράγωνον διε-
λεῖν εἰς δύο τετραγώνους. τετάχθω πάλιν
ἡ τῷ πρώτου πλευρᾶ ζ' ἰνὸς, ἡ ῇ τῷ ἑτέρου
ςϛ̅ ὅσων δήποτε λείψει μ̅ ὅσων ἐστὶ ἡ τῷ διαι-
ρουμένου πλευρᾶ. ἴσω δὴ ςˢ β̅ λείψει μ̅ δ̅.
ἴσονται· οἱ τετράγωνοι ὁ μὲν δυνάμεως μιᾶς,
ὃς δὲ δυνάμεων δ̅ μ̅ ιϛ̅ λείψει ςϛ̅ ιϛ̅. βού-
λομαι τὸς δύο λοιπὸς συντεθέντας ἴσους ᾗ μ̅
ιϛ̅. δυνάμεις ἄρα ε̅ μ̅ ιϛ̅ λείψει ςϛ̅ ιϛ̅. ἴσαι
μ̅ ιϛ̅. κỳ γίνεται ὁ ἀριθμὸς ιϛ̅ πέμπτων.

H iii

OBSERVATIO DOMINI PETRI DE FERMAT.

CVbum autem in duos cubes, aut quadratoquadratum in duos quadratoquadrates & generaliter nullam in infinitum vltra quadratum poteftatem in duos eiuf- dem nominis fas eft diuidere cuius rei demonftrationem mirabilem fane detexi. Hanc marginis exiguitas non caperet.

L'HÔPITAL'S RULE–DIFFERENTIAL CALCULUS

$$\lim_{x \to c} \frac{f(x)}{g(x)} = \lim_{x \to c} \frac{f'(x)}{g'(x)}$$

$$\lim_{x \to c} \frac{f(x)}{g(x)} = l.$$

Guillaume François-Antoine (1661–1704), popularly known as the Marquis de l'Hôpital, was a renowned French mathematician who is credited with the rule of calculating the limiting value of a fraction where the numerator and denominator tend to zero or both tend to infinity.

This rule appeared in his work *Analyse des infiniment petits pour l'intelligence des lignes courbes* (1696), the first text that ever presented differential calculus, although it is known that this rule is thanks to Johann Bernoulli (left), who is believed to have conceived of the proof first. L'Hôpital never claimed to be the author of the rule, indeed, the author is not even mentioned in the published book.

In mathematics, L'Hôpital's rule uses derivatives to help evaluate limits of a nonspecific nature, that is, when we encounter situations of the type 0/0 or ∞/∞. The application of this rule often converts an indeterminate form into a determinate form, thus allowing the limit to be evaluated easier.

If when calculating the limit we find again the conditions established by this rule, the rule can be re-applied as often as we consider appropriate to achieve the sought limit. Yet to resolve the remaining uncertainties, this rule cannot be applied directly. In these cases, they have to be transformed into an uncertainty of type 0/0 or ∞/∞ and then L'Hopital's rule is applied.

Three centuries after the presentation of the theorem, L'Hôpital's rule is probably the most popular mathematical tool used in many scientific fields because it allows approximate values that tend to infinity.

In the introduction to his famous work, published anonymously, L'Hôpital acknowledges the collaboration of Wilhelm Leibniz, Jacob Bernoulli and Johann Bernoulli.

ANALYSE

DES

INFINIMENT PETITS,

Pour l'intelligence des lignes courbes.

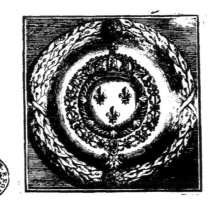

989. **A PARIS,**

DE L'IMPRIMERIE ROYALE.

M. DC. XCVI.

SKIP

FERMAT'S SPIRAL

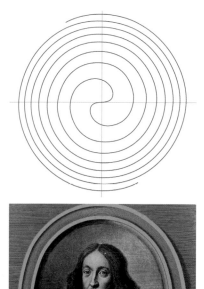

The lawyer Pierre de Fermat (1601–1665) was, along with René Descartes, one of the leading mathematicians of the 17th century. Fermat is credited with the discovery of differential calculus before Newton and Leibniz, he co-founded the theory of probability with Blaise Pascal and he described the fundamental principle of analytic geometry. He is known as the "prince of amateurs," given that he studied and analyzed mathematics in his spare time while maintaining his law practice.

The spiral that bears his name, also known as the parabolic spiral, is a special case of the Archimedean spiral. It is a curve that responds to the equation $r = a + b\theta$ and represents a spiral that diverges from a point and moves away from the same point logarithmically. Fermat's spiral is based on the equation of a spiral in which each value of the angle θ will correspond to two values of r, one positive and one negative, that is, $r = \pm\, \theta^{1/2}$.

From this equation, two symmetrical spirals are formed that complement one inside the other and continue to expand to infinity. The idea is based on the conditions that r is a locus of a point moving at a constant speed on a straight line rotating about a fixed point of origin at constant angular velocity.

Fermat's spiral can be observed in forms of nature, such as in the arrangement of sunflower seeds and the floral disc of the daisy.

MARKOV CHAINS

The probability that an event occurs depends *inter alia* on the preceding event. This statement is known as the Markov chain, named after Andrei Markov the Russian mathematician (1856–1922), who came up with it in 1907.

A Markov chain corresponds to a special type of process that commands a "short memory" in the sense that it only "remembers" the last state of a chain of events to decide the future event. This dependence on past events distinguishes this type of probability from the probability of independent events, such as rolling a die, in which the sequence of events is not influenced by the previous result.

A Markov chain is a system that changes its state over time, so it does not allow an infinite prediction (as would be possible in other circumstances). For example, when flipping a coin an infinite number of times we know that each time it is flipped there is a $1/2$ probability that it is heads and 1/2 that it is tails, so we can say with certainty that for any given time we may flip it, n, the probability that it will be heads will be $1/2$. In a Markov chain, it would be impossible to make a prediction like this, because to know what will happen in the nth event we must already know the outcome of event $n-1$, that is, we need to know the current status and all the probable possibilities.

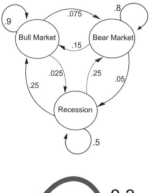

Thus, the dependence between random variables is such that, if we know the value of the event Xn, the values of $X0$, $X1$, $X2$… are irrelevant to study the value of $Xn+1$. This does not mean that $Xn+1$ is independent of $X0$, $X1$…, but these variables have an influence on $Xn+1$ only through Xn, as, for example, in weather forecasting. More precisely, if we call the period $n+1$ future, the period n present, and the periods 0, 1, 2… past, we can say that the past influences the future only through the present.

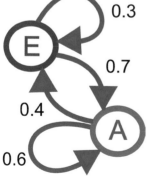

In addition to highlighting the study of probability in mathematics, Andrei Markov was a staunch political activist who opposed the privileges of the Tsarist nobility, to the point of earning the nickname the "militant scholar."

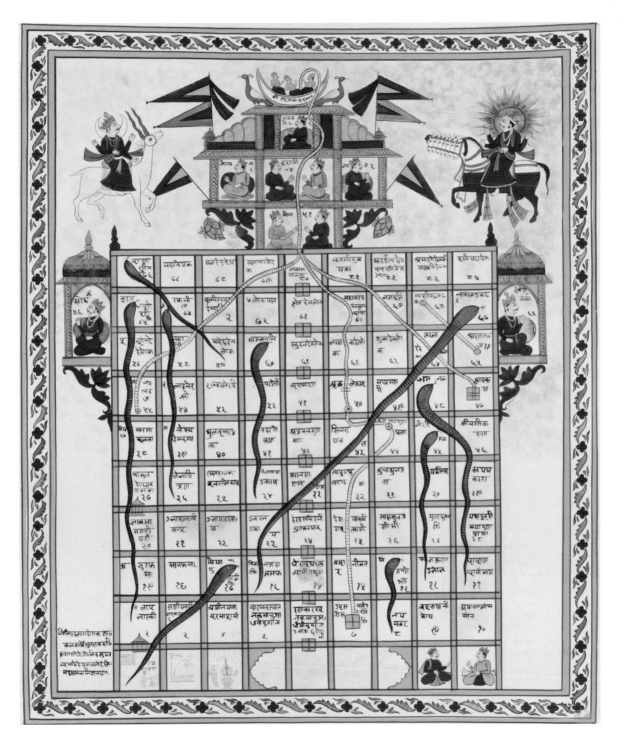

use?

THE POINCARÉ HYPERBOLIC PLANE

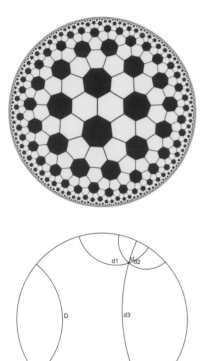

Hyperbolic geometry arose as a geometric model that does not meet all the postulates of classical geometry. Hyperbolic geometry is based on a model of constant and infinite curvature, first proposed by Jules Poincaré (1854–1912), a French mathematician and philosopher, around 1894. The main difference is that it does not satisfy Euclid's fifth postulate on parallel lines.

When we consider the classical Euclidean plane, we are used to the fact that two perpendicular lines to a fixed line *r* are parallel and therefore converge to a point at infinity. So what differs in the Poincaré plane? The Poincaré plane is a disk of constant curvature, so we must imagine it as if we were sitting on a sphere. Now draw, in this sphere or Poincaré plane, a fixed line which we also call *r* and try to draw two lines that intersect *r* at right angles. In this case what we are describing are two circles, which will start and finish at a given point and therefore will not cross at infinity, that is, those perpendicular lines to *r* do not run parallel to each other because they do not run toward the same limit point at infinity.

Here, we might consider that we live in a world and a universe with a clear and indisputable curvature, thus it follows that hyperbolic geometry based on the Poincaré plane is a schema that seems to describe the universe in the best manner possible. But actually the area differences that result from applying Euclidean versus hyperbolic geometry are so small as to be barely noticeable, so that Euclidean geometry is an acceptable approximation for any ordinary scale.

Jules Henri Poincaré was a renowned and prestigious mathematician, theoretical scientist and philosopher of science, often described as the man who almost discovered relativity, as it is said that his numerous works inspired Albert Einstein to do so.

use, NO

THE LEVIATHAN NUMBER

Among the largest numbers that we are able to imagine, we find the Leviathan number, which in mathematical notation is expressed as $(10^{666})!$. In mathematics, the exclamation mark is the factorial operator, which is obtained by the multiplication of all numbers less than and equal to that indicated, for example, the factorial of 5 $(5!)$ would be $5 \times 4 \times 3 \times 2 \times 1 = 120$.

In the Old Testament and therefore in Judeo-Christian literature, the name of Leviathan denotes a colossal sea monster, often associated with the devil or Satan. Tradition relates the devil with the number 666, which mathematically also presents a few peculiarities, such as the sum of the first 36 natural numbers, the sum of the squares of the first seven prime numbers $(2^2 + 3^2 + 5^2 + 7^2 + 11^2 + 13^2 + 17^2 = 666)$, and the sum of the cubes $1^3 + 2^3 + 3^3 + 4^3 + 5^3 + 6^3 + 5^3 + 4^3 + 3^3 + 2^3 + 1^3$.

The Leviathan number is only a curiosity, it is virtually impossible to calculate or to use in calculations, but two characteristics are known for certain about Leviathan's number: the first six digits are 134072, and the total number of digits that comprise it are greater than 10^{668}, and for this reason it is estimated that we would need more than 3×10^{660} years to write out all its figures.

Gustave Doré (1832–1883) was a French artist, and one of his famous works is the engraving Destruction of Leviathan *(1865), which represents the death of this beast at the hands of God.*

MATHEMATICS   **LAKES OF WADA**

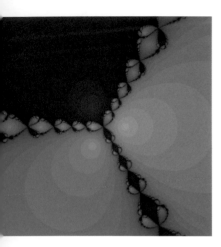

The construction of the lakes of Wada was first described by the Japanese mathematician Kunizō Yoneyama in 1917, who named it in honor of his professor Takeo Wada. It is a mathematical model in which three disjoint connected open sets of a plane share the same boundary.

There will be points along the border which will establish the boundary of two of these regions, and there may even be some point that will establish the boundary of the third. Although intuition makes us predict that the number of points that are shared may be scarce or limited, the model of the lakes of Wada states that this number is infinite. The example by Yoneyama describes this mathematical construction.

Let us imagine an island in the middle of the sea that has a hot lake and a cold lake and on which the following will be done. During the first hour, one digs three channels - one with sea water, another with water coming from the hot lake and the last with water coming from the cold lake - so that the various types of water are never in contact. Within the hour, every point of land is at a distance of less than a mile from each type of water, salt, cold and hot.

In the next half hour, each of the channels grows longer, always avoiding that the different types of water come into contact, and at the end of the work, the distance between each point and each type of water is less than half a mile. The work proceeds in the same manner during the next $1/4$ hour, and the next $1/8$ of an hour, and so on. At the end of the second hour (remember that $1 + 1/2 + 1/4 + 1/8 + ... = 2$), the dry ground forms a closed complex, and close to any of its points there is hot, cold and sea water. This complex is the common boundary to the three regions: the sea, the cold lake and hot lake extended in their respective channels.

Although it is difficult to imagine, three disjoint connected open sets of a plane may share the same boundary. In general, three regions in two dimensions (for example, three countries) can coincide only at one point, but topologically the commonalities are infinite.

A LEGEND ABOUT CHESS

Legend has it that a long time ago, a king named Sheram reigned in parts of India. In one of the battles in which his army fought, he lost his son, and that left him deeply disturbed, and he only thought about what strategy would have avoided him losing his heir.

One day a young man named Sissa appeared in his court and requested a hearing. The king agreed and Sissa presented him with a game that would take him out of his misery and would provide distraction: a large board with 64 squares on which he placed two collections of pieces. Patiently, he taught the king, ministers and courtiers the fundamental rules:

Each player has eight pawns, representing the infantry advancing on the enemy to disband it. Behind come the elephants of war (the towers), and the cavalry, indispensable in combat. To intensify the attack, the noble warriors (bishops) of the king are included. Another piece with a range of movements, more efficient and powerful than the others, represents the patriotic spirit of the people, the Queen. The piece that everyone should protect completes the collection: the king.

The sovereign was fascinated by the new hobby and played several games driving away his sorrow. Grateful for this magnificent gift, he offered Sissa the opportunity to ask for whatever he wished for. Sissa rejected the reward, but the king insisted and he asked for the following:

"I want to plant a grain of wheat to represent the first square of the chessboard, two for the second, four for the third, and so on, doubling the number of grains for each square, and give me the number of grains of the resulting wheat."

The king was surprised by the strange request, believing it was a reward too small for such an important gift, though he accepted.

After a few hours, the king's most competent algebraists informed the ruler of the kingdom that he was not going to be able to fulfill his promise: it would take $1 + 2 + 4 + 8 + ... + 2^{62} + 2^{63}$ grains of wheat, equivalent to 18,446,744,073,709,551,615 grains. In each kilogram, there are approximately 28,220 grains, so that the result would be about 653,676,260,585 tonnes. The whole of India, planting all their fields and destroying all their cities, would not be big enough to produce during a century the number of grains calculated.

The reward that Sissa claims responds to a geometric progression constituted by a sequence of elements in which each is obtained by multiplying the previous one by a constant called the progression factor or ratio.

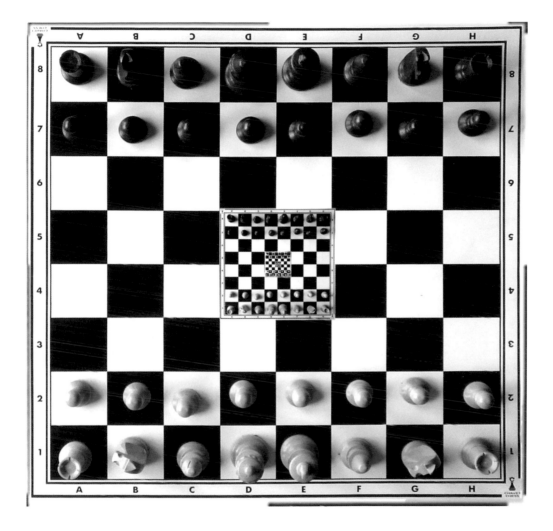

GÖDEL'S INCOMPLETENESS THEOREMS

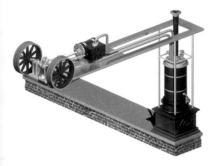

The incompleteness theorems proposed in 1931 by the Austrian Kurt Gödel (1906–1978) are two theorems of mathematical logic that include in their strictly logical mathematical system claims that cannot be proved or disproved from the propositions of the system. Therefore, the basic propositions of arithmetic can lead to inconsistencies. This leaves our system of mathematics essentially incomplete.

To illustrate the theorem, an old paradox is normally used, called the Epimenides or the liar's paradox, which says, "Cretans are all liars." As Epimenides was a Cretan, we can infer that if this statement is true, it means that the statement is false, which contradicts our first hypothesis. On the other hand, if what he says is false, the statement must be true, which brings us back to a contradiction. This line of reasoning is an example of a "strange loop." Any initial assumption that is made leads directly to a refutation of that assumption. Many of the artistic optical illusions of M. C. Escher are based on this concept.

Mathematicians and philosophers from the time of Gödel believed these paradoxes to be an artifact of language that could not be reproduced with mathematics because math is pure and rational. However, Gödel formulated a few mathematical equivalents of this paradox. His theorems could be summarized with the phrase "this statement is unprovable," and in the end, the theorems were impossible to demonstrate as such. What the theorems did demonstrate were the limits of mathematical theory in general.

At that time, attitudes toward mathematics were so optimistic and credulous that it was believed math could decode all of the underlying systems and mechanisms at work in the universe, and the truth or falsity of infinite propositions. The publication of Gödel's theorems dealt a heavy blow, as they showed that humanity could not reach absolute knowledge and truth through the study of math alone.

Kurt Gödel proved a fundamental theorem that caused a stir in mathematical circles, that there are mathematical statements that can neither be proved nor disproved.

usr?

KOCH SNOWFLAKE

In 1904, Swedish mathematician Helge von Koch (1870–1924) described a famous fractal curve in an article entitled *On a continuous curve without tangents, constructible from elementary geometry*. This curve provides the basis for the construction of another fractal object, the Koch snowflake, one of the simplest fractal figures and one of the first to be described, similar to the curve except that it starts from a triangle instead of a segment.

The fractal is constructed by an iterative process that begins with an equilateral triangle in which ultimately each one of its sides is replaced by what is called the Koch curve. To draw a Koch snowflake, a segment is divided into three equal parts and an equilateral triangle is built on the middle part. Then the base of the triangle is removed and you are left with four segments. Proceed in the same way with these four segments, leading to 16 smaller segments in the second iteration, and so on. If instead of starting with a segment, we start with an equilateral triangle, we obtain the Koch snowflake.

By repeating the procedure in each resulting segment theoretically to infinity, the perimeter of the snowflake tends to infinity. You will arrive at the paradoxical case of an infinite perimeter enclosing a finite area (which tends to $8/5$ of the value of the initial area).

Fractals exhibit self-similarity, the property of an object in which the whole figure is similar to a part of itself, that is, the same configuration is repeated as much as we expand the figure's size.

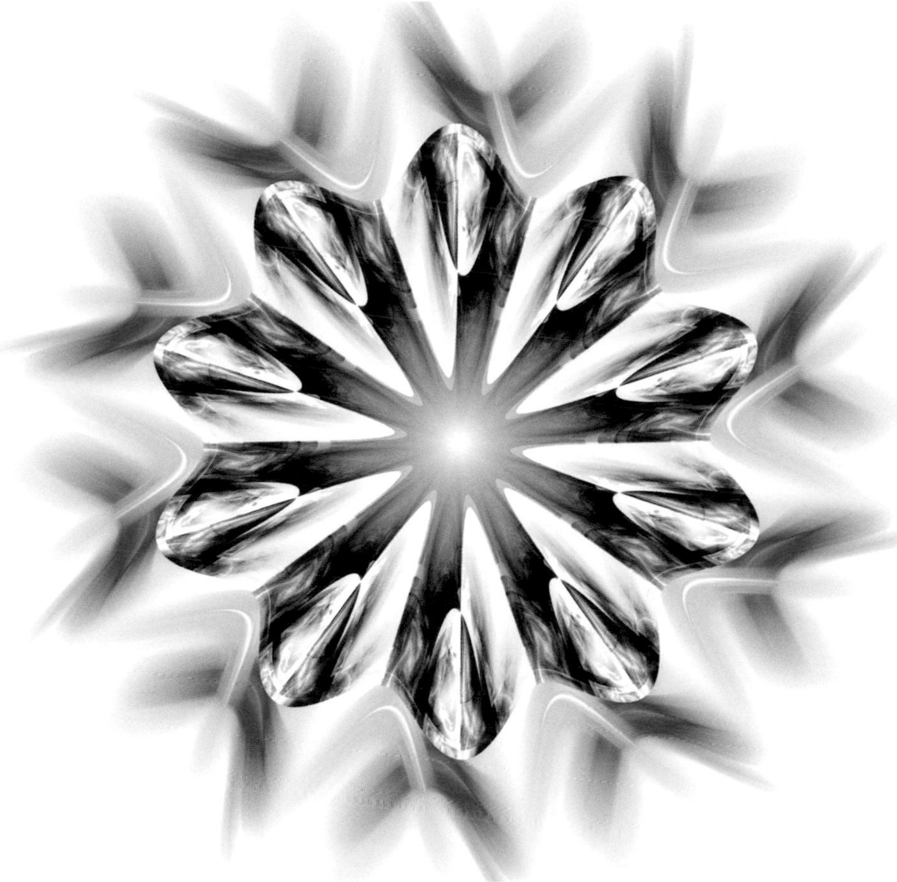

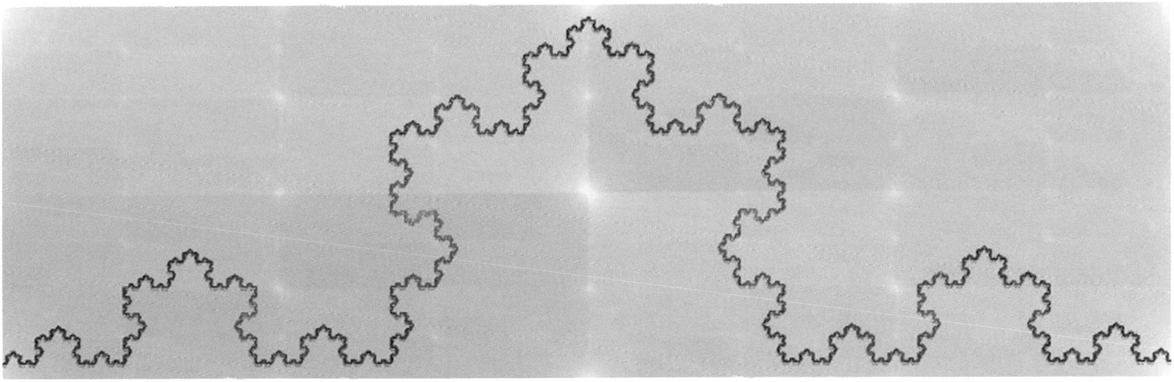

TECHNOLOGY

was?

STORAGE OF SOLAR ENERGY

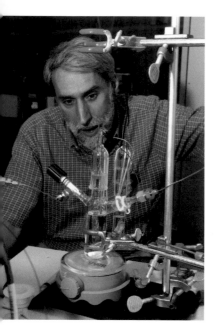

One of the major disadvantages of solar energy is that when the sun does not shine, no electricity is harnessed, and when it doesn't stop shining, excess energy cannot be stored for later use in times of scarcity. Rather, storage of excess solar energy can be done, but the methods are so expensive and inefficient that they are prohibitive.

A new discovery by Daniel Nocera and Matthew Kanan could revolutionize the use of renewable energy. These engineers at the Massachusetts Institute of Technology have managed to store the sun's energy through a simple, highly efficient and affordable process. What's more, it only requires natural, non-toxic materials.

The scientists were inspired by nature, by the photosynthesis of plants, to develop an unprecedented process that will allow solar energy to be used to divide water into hydrogen and oxygen. These gases can then be recombined inside a fuel cell, creating electricity that can power a house or an electric car, day and night, without emitting greenhouse gases.

The key component in this process is a new catalyst that produces oxygen from water and another catalyst that produces hydrogen. The catalyst consists of cobalt, phosphate and an electrode placed in the water. When the electricity, whether from a solar panel, a wind turbine or any other source, passes through the electrode, the cobalt and phosphate form a thin film on the electrode and oxygen is produced. Combined with another catalyst, such as platinum, that can produce hydrogen from water, the system can mimic the separation that occurs during photosynthesis, which divides water into its two components, oxygen and hydrogen. The new catalyst works at room temperature with plain water.

Solar energy has incredible potential: the energy that we harness from all of the world's coal, oil and natural gas reserves can be compared to only 20 days of sunlight.

Producer of an infinite fuel source?

FUEL FROM PHOTOSYNTHESIS

An American biotech firm, Joule Unlimited, is developing a method of producing fuels that can revolutionize renewable energy research: it has patented a genetically altered bacteria that produces biofuel through photosynthesis, that is, using sunlight and consuming carbon dioxide.

It is a type of cyanobacteria, also known as blue-green algae, although they are not exactly algae. These bacteria are characterized by performing a special type of photosynthesis, known as oxygenation, which combines carbon dioxide with water to form sugars from which they feed. In this way, they accumulate a lot of oxygen, which they then release. The emergence of these bacteria in the evolutionary history of our planet is the reason that we have so much oxygen today.

Joule Unlimited has modified a cyanobacterium through bioengineering so that, through this photosynthesis, oxygen is obtained not only as a byproduct, but also secretes molecules called alkanes, hydrocarbons chemically indistinguishable from those produced artificially in an oil refinery. These hydrocarbons are the main ingredients of diesel fuel.

Alternative energy experts agree that photosynthesis is a promising avenue in terms of biofuels research. One stumbling block is the need to convert the resulting product of this natural process into fuel. Many companies are trying to develop an algae that performs this process, but energy is required to separate the algae from the water and then process the generated oils and that convert it into fuel. An organism that secretes the desired product directly avoids both problems.

Joule Unlimited's bacteria need only sunlight, carbon dioxide and water. They can grow in water unsuitable for consumption and use wastelands for cultivation, which makes them a serious contender over other biofuels such as corn or sugar cane, which tend to use land previously reserved for forests or agriculture.

The potential of microalgal photosynthesis has already been demonstrated on more than one occasion in the history of the planet: these organisms filled the atmosphere with oxygen and the sea with carbon dioxide.

use Infinite operation? of clock (handwritten annotation)

TECHNOLOGY

THE ATMOS CLOCK

Jaeger–LeCoultre is a brand of luxury watches based in Le Sentier (Switzerland). In addition, it also has a long tradition as a supplier of parts and tools to the Swiss watch industry.

Talented inventor and self-taught watchmaker Antoine LeCoultre (1803–1881) founded a small workshop in 1833 to manufacture high quality watches in Le Sentier. In 1844, he invented the most precise measuring instrument, the *millionomètre*, the first instrument capable of measuring a micron, a millionth of a meter. This meant that components could be produced with rigorous accuracy, resulting in more accurate chronometers. Three years later, LeCoultre & Co. created the first self-winding pocket watch with a string system and time-setting mechanism using the crown. This revolutionary breakthrough left key wind-ups obsolete. In 1851, LeCoultre was awarded a gold medal for his work in precision watchmaking and mechanization in the first Universal Exhibition in London. In 1903, with his son Elie in charge of the company, LeCoultre & Co. formed an association with the Parisian watchmaker Edmond Jaeger (1858–1922). True horological marvels were born out of this relationship, which led them to establish a brand in 1937.

Invented in 1926 (and patented in 1928) by engineer Jean–Léon Reutter, and perfected and manufactured by Jaeger–LeCoultre, the Atmos clock is a mechanical watch with perpetual movement. No need for string, battery or any other external force, its infinite mechanism obtains power to operate from the slightest variations in temperature and atmospheric pressure, which causes contraction or dilation of the gas in a sealed bellows that expands and contracts, and for this reason it seems as if the clock "breathes." A change of one degree Celsius allows the Atmos to operate for 2 days. It is so efficient that even today it is still the mechanism that consumes the least amount of energy in the world.

A conceptual and ingenuous basic operation along with a flawless aesthetic has converted Atmos clocks into perfect gems of art.

Use Telomeres Infinite Life

TA–65, LIVING INFINITELY

In genetics it is considered that telomeres, the ends of chromosomes that protect them from deterioration, are "the aging clock" of the human body. Many scientists believe that life expectancy and declining health is imposed by the progressive shortening of our telomeres. As a normal cell divides, it loses fragments of telomeres, which causes a progressive decline in function and ultimately cell death. Repeated studies agree that, were it not for the progressive deterioration of telomeres, human cells could divide infinitely and therefore could be immortal.

This aim could be achieved by increasing levels of telomerase. Telomerase is an enzyme, that is, a protein that catalyzes metabolic reactions and repairs and lengthens telomeres, whose function is the structural stability of chromosomes and directly affects the rate at which cells age. The activation of this enzyme could prevent the telomeres from shortening, which would slow or even halt the aging process.

Recently, researchers at Sierra Sciences, in collaboration with TA Sciences, Geron Corporation, PhysioAge and the Spanish National Cancer Research Center, have discovered TA–65, an amazing natural compound capable of activating the gene for telomerase and lengthening telomeres of human cells, which may make it possible to stop or even reverse the aging process, bringing us closer to the myth of eternal youth.

Telomeres were identified by the American biologist Hermann Muller in the thirties. Since then, we've deepened our understanding of these structures that are closely related to aging and cancer.

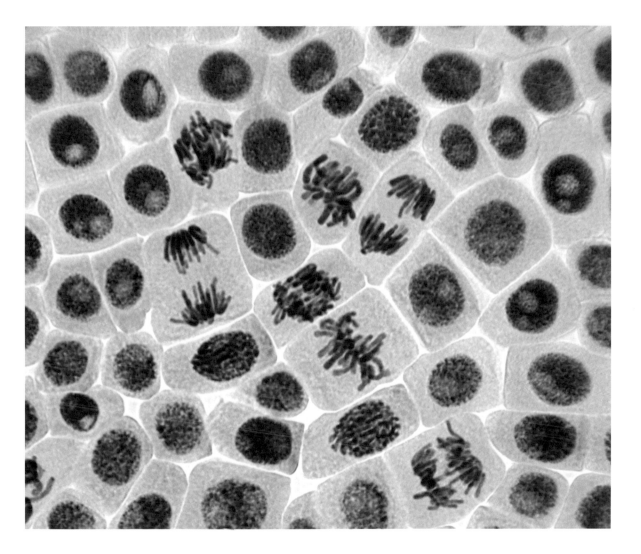

INFINITE LOOP PROGRAMMING

In programming, a loop or cycle is a type of control structure to repeat one or more statements (lines of code) multiple times. Loops are executed when a certain condition is met, or, in computer language, while that condition is true. Generally, a loop is used to carry out an action repeatedly without having to write the same code several times, which saves time and makes the code clearer and easier to change in the future.

The loop is an evolution of the assembler code, where the only possibility of iterating a code was to establish a "jump" (JMP) sentence, which in high level programming languages was replaced by the "go to" (GOTO). The three most commonly used are the "while" loop, the "for" loop and "repeat" loop.

In most programming languages, you can also "escape" or "break" a loop, at which point, although the condition remains true, another way out of the execution of the loop is embedded. The use of these functions is considered improper and unnecessary by the most purist programmers, as you can always exit a loop by using the end condition. When exiting is difficult or impossible, it means the wrong type of loop has been chosen. But from a practical standpoint, it is often easier to use a loop escape function.

In the case of infinite loops, the end condition is never met, so it continues to run indefinitely, which also occurs if there are any errors in the code. In the latter case it is considered a programming error, unless an indefinite run is the result sought by the programmer, as in the case of malware, so that the action is repeated over and over again.

Malware infiltrates a computer without the owner's consent and damages or extracts the stored data. The term is widely used among professionals to refer also to programming errors that produce such damage.

Use? infinite pen

THE INKLESS METAL PEN

Urban legend says that during the space race, NASA spent an enormous amount of money developing a pen that could operate in zero gravity space, while the Russians chose to carry pencils in their spaceships. The kernel of truth in this story is that an alternative to pencils was sought as the tips of the pencils contain graphite, which is a conductive material. It was felt that broken graphite tips could float around the spaceship and come into contact with a circuit.

Decades later the instrument that would have solved the famous problem exists: The Inkless Metal Pen. This stainless steel or aluminum pen, depending on the version, has a lead tip that, as you write, deposits tiny amounts of this metal onto the page following the lines you trace. Enough particles remain on the paper so you can see the writing, but not enough for the density of the bar of lead to wear out, so it is able to last almost forever without being replaced. This does not happen with graphite or ink, materials subject to rapid wear that have to be replaced in the form of refills or new leads.

Although its appearance is more like a pencil, the Inkless Metal Pen shares properties of pens, because its writing is indelible. It does not stain or leak and it can even be used under water, and if you ever find yourself in zero gravity, you could still use this pen.

Although it may seem unique, the ancient Egyptians, Greeks and Romans used small disks of lead to draw lines on papyrus, and towards the 14th century European painters used rods of this material to create pale gray drawings.

use — infinite filament for light bulb

THE BULB THAT NEVER GOES OUT

One single bulb has been lighting up a fire station in Livermore, California for more than 100 years. It is the most persistent, tenacious and inexhaustible light bulb known to history, the only one with a Guinness record.

It is a 60 watt bulb (although today its power does not exceed 4) with a hand-blown carbon filament that has not stopped shining since it was fitted on June 1, 1901. Since then it has maintained its goal of lighting up the station, and it has only taken a break with the occasional blackout and during a move. It has accumulated more than one million hours of service.

The Livermore bulb is also one of the first examples of a technology in extinction because bulbs of this model are now being replaced with more efficient energy solutions such as OLED lamps or low consumption light bulbs.

Invented by the French scientist Adolphe Chaillet, this bulb was manufactured by the now defunct Shelby Electric Company. Its carbon filament is perfectly isolated in the glass ampoule and operates in a vacuum, as opposed to in a gas-filled space, as is the case with current bulbs. As the bulb remains continuously lit, it does not endure the standard wear and tear of being turned on and off, actions that result in stress cracks and eventual burnout.

Debora Katz, a physicist at the U.S. Naval Academy, has thoroughly studied the properties of the light bulb of Livermore. To do this, and given the impossibility of studying the original bulb without turning it off, the researcher obtained a similar one (burnt out), also manufactured by the Shelby Electric Company in the late 19th century. Katz concludes that the Livermore bulb is different from a modern incandescent light bulb in two ways. First, the filament is about eight times thicker than modern bulbs. Second, the filament is a semiconductor. When a conductor overheats, its ability to conduct electricity fails. However, as the Shelby light bulb warms up, it becomes an even better conductor.

General Electric bought the patent for this unique bulb in 1912 but stopped making it after 2 years to launch one of wolfram, which shone brighter and in theory was of better quality.

use?

TECHNOLOGY

THE ITER NUCLEAR FUSION REACTOR

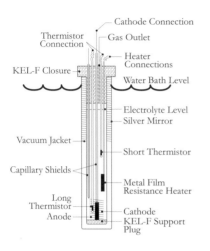

In Cadarache, France, a reactor is being built that attempts to reproduce in a controlled manner how thermonuclear energy is generated in the solar core. It may be a vital source of alternative energy of tomorrow, clean and almost inexhaustible, the realization of which would mean the end to the era of fossil fuels.

In the 1950s, scientists Igor Tamm and Andréi Sájarov, following the original idea of Oleg Lavrentiev, invented the *tokamak*, a Russian acronym for "toroidal chamber with magnetic coils." The tokamak is a fusion reactor that has a toroidal vacuum chamber (which resembles a donut). Confined gas flows on the inside in the state of plasma at over 150 million degrees Celsius, which is guided from the exterior by a magnetic field so that the plasma does not touch the walls of the chamber and temperature is lost.

After building several experimental tokamaks, in 1968 Russian scientists were able to induce a series of thermonuclear explosions resulting from the merger of atoms without producing anything but a stationary power. That was the first step toward a new source of natural and powerful energy, because instead of breaking the nuclei of atoms, as existing nuclear power plants do, different atoms joined together, which was like imitating the behavior of the Sun.

The technology used in a tokamak is that used in the international thermonuclear experimental reactor. The ITER is based on the nuclear fusion of hydrogen atoms. Its fuel is a mixture of deuterium and tritium, two isotopes of hydrogen. There are a few different reasons for this choice: the lighter nucleus is easier to merge. It is also an inexhaustible fuel: deuterium is found in large quantities in water, while the tritium is produced in the same fusion reaction.

ITER will be a tokamak capable of generating 500 million watts of fusion power continuously for 10 minutes. It is scheduled to come into operation for the first time in November 2019. Its life expectancy is 20 years, and then it will be deactivated and dismantled, a phase that can take up to 40 years. If successful, there would still be a long way to go before the arrival of commercial reactors, which is not expected before 2050.

The European Union, China, USA, India, Japan, South Korea and Russia are participants in the International Thermonuclear Experimental Reactor (ITER) project.

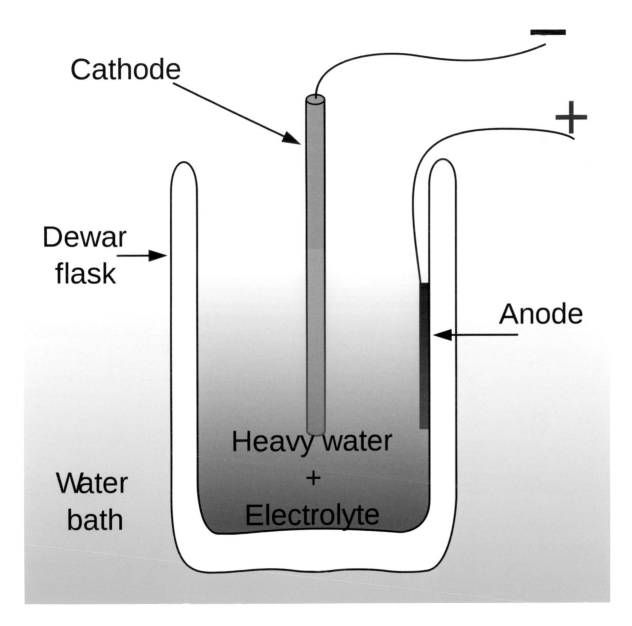

WIND ENERGY

Wind energy is a type of renewable energy powered by the force of the wind. Since ancient times, it has been used to move boats powered by sail or operate the mill machinery to move their blades, though its use to generate electricity is more recent. Currently, the typical way to harness this energy is through the use of wind turbines.

Although Poul La Cour is considered to be the progenitor of modern wind turbines, building his first in 1891, Charles F. Brush had already built a 12 kW wind turbine years prior, and had stored the energy in 12 batteries. The first AC wind turbines emerged in the 1950s thanks to the Dutchman Johannes Jull, who also designed a wind turbine that changed direction depending on wind direction.

The operation of wind turbines is very simple: the wind moves the blades of the rotor connected to a generator through a series of gears, transforming the rotational mechanical energy into electrical energy. A safety brake locks the rotor when the wind speed exceeds a certain value. Wind turbines can be classified according to the orientation of the rotor shaft. On the one hand, there are the original turbine models, those of vertical axis, whose blades rotate in a plane parallel to the ground. On the other hand are those with a horizontal axis, the most widely used model, whose blades rotate in a plane perpendicular to the ground. Wind turbines can work alone or in wind farms inland or on the coast, and can even be installed in bodies of water up to a certain distance from the coast.

The largest wind turbine installed to date is the E-126 by the German company Enercon. At 454 feet (138 m) tall it is an enormous structure; 6 megawatts per year can be harnessed, enough energy to meet the needs of 5,000 households.

Aeolian energy takes its name from Aeolus, the god of wind in Greek mythology. The implementation of wind energy systems is especially popular in Germany, Spain and Denmark.

THE LARGE HADRON COLLIDER

Ironically, we had to build the biggest and most powerful machine in the world to detect the smallest particles that make up the universe. On the French-Swiss border, buried at a depth ranging from 160 to 574 feet (50 to 175 m) underground, lies the Large Hadron Collider (LHC), which aims to recreate the very moment of the origin of the universe.

Inside the 16 1/2-mile (27 km) circumference tunnel, two beams of protons circulate in opposite directions. When they almost reach the speed of light, they will collide head-on, converting their energy into a mass of new particles bringing to life the famous Einstein formula, $E = mc^2$. These collisions will be closely studied in the heart of four large detectors, and from this information, the primordial conditions of energy, temperature and matter will be recreated as they existed when the universe was less than a trillionth of a second old.

The main objective is to reveal the existence of infinitely small particles, such as Higgs boson, described in 1964 by the British physicist Peter Higgs as responsible for giving mass to matter after the Big Bang, which made the formation of the cosmos possible. Proof of its existence would be a great discovery in physics and confirm the theories of the so-called standard model of particle physics, which explains the fundamental interactions between elementary particles. In July of 2012, the discovery of a particle with properties consistent with the Higgs boson was announced. Further research supports the theory that the particle in question is indeed the Higgs boson.

The Higgs boson is not the only possible discovery to be made. Along with this particle, the LHC is also trying to unravel the mysteries of the Big Bang, to determine whether there are more than three dimensions in our universe and to understand the differences between matter and antimatter.

A mini-Big Bang, caused by the collision of ions, has successfully been achieved at the LHC, leaving a mountain of scientific data that may take decards to decipher.

Although the Higgs boson is popularly known as the "God particle," scientists agree that it would be more appropriate to call it the "damned particle" due to the effort that was necessary to find it.

TECHNOLOGY   **SOLAR ROADWAYS**

The Sun represents the most important energy source for our planet. If we could harness 100 percent of the light from the Sun that reaches Earth for only 1 hour, we would be able to meet the energy needs of the entire planet for a whole year. The problem that prevents us from doing this is that the harnessing of solar energy by the methods we have available today is limited, and current storage systems are inefficient and expensive.

In the United States, a prototype has been developed, which aims to pave the road network in advanced solar panels, making them a large energy supply network. The project is led by the company Solar Roadways, which aims to replace the asphalt with this innovative solar technology.

The prototype has three sections. The surface is made of a high strength, translucent material to withstand the weight of vehicles and has a texture similar to asphalt to grip the tires. It is also self cleaning and is heated, resolving any circulatory problems resulting from snow and ice. Photovoltaic cells are located in the middle layer to capture the solar energy, and light emitting diodes will display the typical signs and warning messages in the style of the existing vertical panels. Below, the third layer serves to distribute energy and also stores the fiber optic cables for communications.

These panels can last for 20 years of prolonged use, even enduring the heavy vehicle traffic without sinking or suffering internal damages. It is estimated that, for every 4 hours of light per day per mile of a four-lane solar highway, enough energy will be stored to power more than 400 homes.

In the United States, the construction of a test section of about 7 miles (11 km) in Idaho has already been approved, while in the Netherlands the construction of over 300 feet (100 m) of photovoltaic road is being implemented shortly.

? use

RUBIK'S CUBE

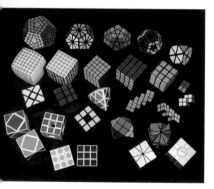

The *par excellence* toy for fans of challenges, mathematics and geometry is a puzzle consisting of a 3 × 3 × 3 cube of colored squares that can be rotated by sections without disassembling the cube, changing the relative position of each square. The aim is to restore the original layout, wherein each facet of the cube is composed of squares of a single color.

This popular puzzle offers a true experience of vertigo, a small glimpse of the infinite. The player must be able to find the only solution within a vast gulf of possibilities. A Rubik's Cube can be arranged in over 43 trillion different permutations, an almost unimaginable number. If we consider that of those 43 trillion positions only one corresponds to the solution of this fabulous and surprising mathematical artifact, we come to the immediate conclusion that trying to solve it randomly by rotating its pieces is a task, and while it may be possible, it is definitely unlikely.

Invented in 1974 by architect, sculptor and professor Enrö Rubik in order to create a tool to explain some geometric concepts, the "Magic Cube" was a piece divided into different parts moving around a central axis, with the aim of, far from a form of entertainment, helping his students at the School of Architecture to acquire a better visual perspective in three dimensions. Marketed since 1980 under its current name, it is estimated that there are over 400 million of these cubes around the world.

In addition to the traditional 3 × 3 × 3 Rubik's cube there are different versions: 4 × 4 × 4, known as "Rubik's revenge," 5 × 5 × 5, called the "Professor's cube" and other geometric shapes such as the pyramid, octahedron, dodecahedron and so on.

TECHNOLOGY   **NANOTECHNOLOGY**

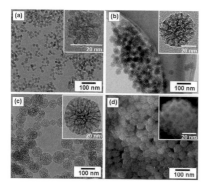

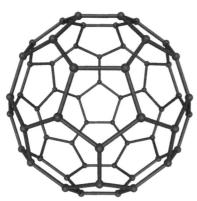

The concept of nanotechnology encompasses the fields of science and technology in which we study, obtain and manipulate very small quantities of materials, substances and devices in a controlled manner, generally less than one micron, that is, studies at the nanoscale.

"Nano" is a Greek prefix denoting a factor of 10 to the power of −9. A nanometer is one billionth of a meter, or, one millionth of a millimeter. Thus, nanotechnology offers the possibility of manufacturing materials and equipment by exploiting the rearrangement of atoms and molecules.

Beyond its concept, we are interested in its potential within the body of current research and applications. Nanotechnology, which arose in the middle of the last century from the Nobel Prize winner in Physics, Richard Feynman, is now poised to launch a second industrial revolution in the 21st century, but its impact on modern life still seems like something from science fiction.

Chemistry, biology and physics are some of the application fields of nanotechnology, which may represent a solution to various problems. Nanotechnology, in addition to the valuable insights it may provide at a theoretical level, is already precipitating innovations in practical fields such as medicine and environmental management.

Thanks to the many nanotechnological advances, today it is possible to modify the structure of molecules from synthetic polymers to proteins with specific physiological functions, create drugs that work at the atomic level, microchips capable of performing complex genetic analysis, generate inexhaustible energy sources, and construct buildings with microrobots, among other applications. Nanotechnology will result in numerous advances in many industries and new materials with extraordinary properties, new applications with incredibly fast components and even biosensors capable of detecting and destroying cancer cells in the most delicate parts of the human body, such as the brain. The possibilities that arise are endless, and many advances in nanoscience will be among the great technological breakthroughs that will change the furure.

Nanotechnology is an interdisciplinary field devoted to the control and manipulation of matter at a submicron scale, that is, at the level of atoms and molecules.

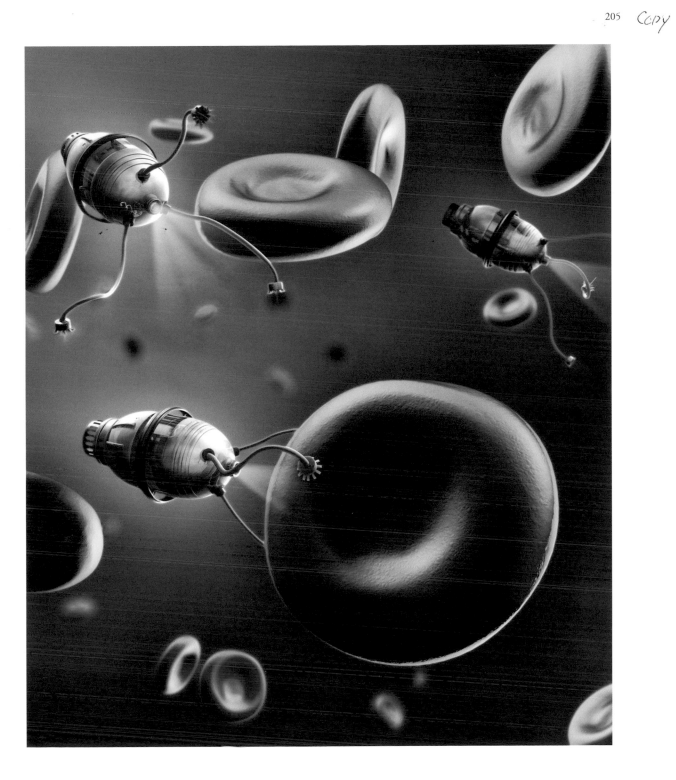

NO

QR CODES

Without a doubt you have already come across these codes in several stores, ads and magazines. A QR code (quick response barcode) is an information storage system using an array of squares and two-dimensional points. It is characterized by its square shape and the three boxes located in the upper and lower left corners.

At the time of their creation in 1994, QR codes were used by administration and inventory management in a variety of industries. Since their inclusion in numerous software programs and mobile phones, they have become popular very quickly and they have many uses, primarily in communications and marketing. Every day an unimaginable number of QR codes are created, which makes us wonder if there is the possibility that they are never-ending.

The number of possible codes is the result of raising the number of possible states of each pixel (white or black) to the number of pixels that the QR in question has. For example, if we consider a QR of 25 pixels high by 25 pixels wide, the code would have a total of 625 pixels. From this we must subtract the number of pixels of position, alignment and synchronization, which helps the camera to recognize the code and be able to interpret it. Some extra pixels are also added to the QR to include information on the version used or the format (numeric or alphanumeric) and even a bug fix, but this number is variable. In total, we will say that on average there are about 250 fixed pixels, leaving us with 375 pixels which can freely switch between two states. The result of possible combinations would be $2^{375} = 7.657 \times 10^{112}$. That would be enough for everyone on Earth to generate 7 billion QR, and there would still be the possibility of creating codes of 29×29, 30×30...

Information encrypted with a QR code can be an image, text, phone number, an electronic flight ticket or a website address: the possibilities are almost infinite.

η_o

USB MEMORY WITH INFINITE CAPACITY

USB memory sticks have become ubiquitous in recent years, as a safe and convenient tool for storing and transporting all sorts of data. Today technology is being implemented that makes the storage capacity of USB sticks possibly infinite.

The possibility that one of these devices could possess infinite storage capacity at first seems completely unreal, but incredibly it is true. This USB drive establishes a Wi-Fi connection between a computer, where all the information that the user wants to transfer or save is saved, and any nearby device with a USB port. Thus, the USB "tricks" the device with memory constraints to which it is connected (such as game consoles, other computers, cameras, printers, televisions) into thinking that it is connected to a hard disk or USB with the stored information. Thus, the system is able to transfer all the data desired with unlimited capacity.

The special feature of this new device is its wireless ability to store and transfer information from one place to another, without requiring physical storage on the USB memory itself. The invention facilitates transmissions between the one device and virtually any peripheral that has a USB port, so that the user transfers files from one to another through a virtual hard disk whose only limitation is the capacity of the underlying computer system.

As a security measure, manufacturers ensure that the wirelessly transmitted information is encrypted, to lower the risk of losing the USB stick with the saved information, as the drive does not save any data. Undoubtedly, it has a vast advantage over the older memory devices, posing a potential solution to current storage and data transfer problems.

USB flash drives were substitutes for floppy disks and CDs. Normally they have a memory capacity of 4, 8, 16 or 32 gigabytes, compared to the 650 or 700 megabytes of a standard CD and 1.44 megabytes of the 3 ½ inch floppy disks.

ATOMIC PRECISION

The concept of atomic precision dates back to 1879, when Lord Kelvin suggested the use of atomic vibrations to measure time. The first atomic clock that used magnetic resonance was built in the 1930s.

Atomic clocks achieve their renowned accuracy due to their mechanism, capable of measuring in excruciating detail natural vibrations due to magnetic resonance of certain atoms and molecules. The resonance entails that each chemical element and compound absorbs and emits electromagnetic radiation at its own characteristic frequencies (the number of oscillations of the electromagnetic wave per unit of time). These resonances are stable across space-time, that is, a hydrogen atom registers the same frequency today as it would have a million years ago, and as it would even in another galaxy.

One of the most common elements for the construction of atomic clocks is cesium, and all atomic clocks used today in the world are based on its physical properties, because it yields a margin of error of less than 1 second every 150 million years. In 1967 the reliability of atomic clocks with cesium forced the International Bureau of Weights and Measures to choose the atomic vibration frequency of these devices as a standard basis for the definition of the physical time unit, establishing that a second corresponds to 9,192,631,770 cycles of radiation associated with the hyperfine transition between different states of a cesium-133 isotope.

Atomic clocks maintain a continuous and stable time scale, the International Atomic Time (TAI). For everyday use, another time scale is used, Coordinated Universal Time (UTC), which is derived from TAI. International Atomic Time and Coordinated Universal Time are used as worldwide time scales for processes such as global communication, satellite navigation and topography, and time stamping of transactions between financial and stock markets.

William Thomson, 1st Baron Kelvin (1824–1907) was a British mathematician and physicist noted for his work in the field of thermodynamics and electronics, and the development of the Kelvin temperature scale.

No

TISSUE ENGINEERING

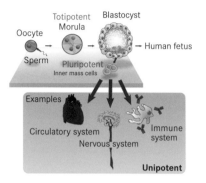

Tissue engineering is a field of research that aims to develop substitutes for organs and tissues damaged by disease or accidents, avoiding transplants that require a human donor.

Currently, tissue engineering opens the door to countless possibilities in terms of tissue and cell regeneration and organ transplants. This is possible thanks to the existence, in both adult and embryonic organisms, of a group of cells known for their ability to change and morph into any cell type, making their use possible in organ regeneration and in the replacement of tissues in general.

There are two main groups of these pluripotent cells: embryonic stem cells (present in embryos and newborns' umbilical cord), which have the capacity to differentiate into any organ and tissue, versus adult stem cells (present in spinal bone in adults), which are able to take a step back in the path of differentiation to form another tissue or cell type. Of these two groups of cells, the latter are the most commonly used because of the ethical implications of using embryonic cells.

The endless possibilities of this field of study have thus far spawned hundreds of clinical trials around the world, most notably the treatment of neuronal damage, the regeneration of cardiac cells, epithelial, cornea and articulate tissue. The most widely used and effective is the autologous bone marrow transplantation. However, much remains to be done, and many studies are underway related to the use of these cells to fight the growth of tumors and metastatic cancers in general.

By controlling and manipulating the cell cycle and its reversal, scientists are able to change undifferentiated cells (A) to differentiated cells such as those of nerve tissue (B) and vice versa.

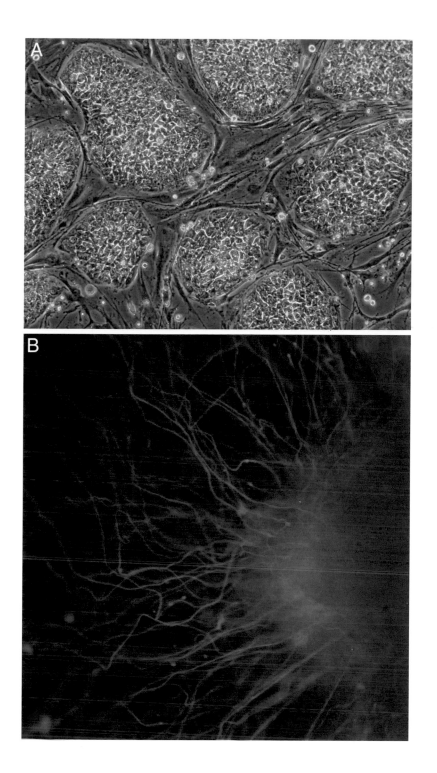

TECHNOLOGY

ISAAC ASIMOV AND THE GROUP MIND

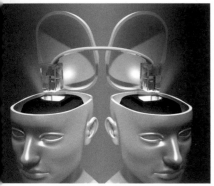

Isaac Asimov (1920–1992) was a writer and biochemist known for being a successful and prolific author of science fiction, history and popular science. Along with Robert A. Heinlein and Arthur C. Clarke, he is widely considered one of the three best writers of science fiction, and he authored some of the best books of the genre. In the framework of science fiction, he was the creator of the three laws of robotics.

In his works, he predicted the operation of a virtual universe in which we would all have a permanent connection to the vast libraries of knowledge and in which any question would have an answer, thus freeing us from the time-consuming task of memorizing vast amounts of information. This source of pervasive knowledge and superior intelligence that solved almost any question was dubbed the Multivac, a fictional computer that appears in many of the stories or tales by Isaac Asimov published between 1955 and 1975, such as *The Last Question*, *All the Troubles of the World* and *The Machine that Won the War*. In Asimov's universe, it is represented as an enormous supercomputer located deep in the underground facilities as a security measure. Like most of the technology that the scientist describes in his fiction, the Multivac's specifications vary in their different appearances.

Asimov died in 1992, prior to the popularization and exponential expansion of the Internet. Today it is often suggested that the Internet has become an all but fully realized manifestation of Asimov's Multivac, since the Internet possesses many of its features, from search engines to integrated services to its seemingly limitless capacity to expand and disseminate human knowledge.

Asimov believed that computers would change the lives of human beings in many ways. In the The Last Question, *he imagined super-computers which were aware of their own existence and which, by using multiple questions and calculations, could answer any question.*

GRAPHENE

Graphene is one of many carbon derived structures, such as charcoal, graphite and diamond. They all share a basic substance, the carbon atom, yet they are chemically arranged in different ways. In fact, graphite from pencil lead is merely the superposition of many layers of graphene.

Graphene is structured as a crystal lattice composed of hexagons in which, at each vertex, there is a carbon atom with links to other atoms, forming a sort of honeycomb structure. A detailed study of the structure has been carried out by physicists André Geim and Konstantine Novoselov, awarded the Nobel Prize in Physics in 2010 for their work on this versatile material.

Graphene is almost 1,000,000 times thinner than the finest hair and is so light that a layer of 1 square meter would weigh less than 1 milligram. It has a mechanical strength 100 times greater than most resistant steels, so this single 1 square-meter layer could maintain a weight of 4 kilograms without breaking. As it is a lightweight material and almost inde-structible, and it is harder than diamond, graphene and its compounds could revolutionize reinforcement material use, aerospace technology, and the automotive industry, to name but a few examples.

With an electrical conductivity 100 times greater than that of silicon and an extraordinary thermal conductivity (so that it does not give off heat), this material may also change the electronics industry. Due to its ability to dissipate heat, graphene processors would be smaller than the current silicon chips. In fact, graphene transistor prototypes are already being manufactured, which operate at a frequency of 100 GHz, 10 times greater than current transistors. As it is transparent, clear electrodes may be manufactured, optimal for the design of flexible solar cells, roller screens, and so on.

A list could be prepared of the many properties and applications of this new material that will be well represented in science and technology in the future.

Geim and Novoselov, both with PhDs in Physics, currently work as researchers at the University of Manchester, where they continue exploring the many applications of this amazing compound.

TECHNOLOGY

THE SOUND BARRIER

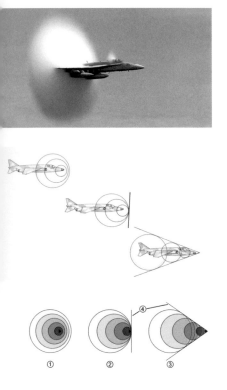

Sound is a wave, and as such, its speed depends on the density and temperature of the medium in which it moves (air, water or solid objects). Through air and at a temperature of 20°C, sound travels at an average speed of 1,125 feet per second (343 m/s). The term "sound barrier" was coined at the end of World War II to describe what was thought to be a physical limit that prevented large objects to move at supersonic speed.

When traveling at speeds approaching that of sound, many prototypes suffered a loss of pressure and aerodynamic problems. That is because when approaching the speed of sound, the air flowing over the surface of the aircraft compresses and produces a considerable increase in the resistance, resistance to which the aircraft is not designed. Previously it was thought that this resistance was growing exponentially as the speed would increase, preventing the crafts from reaching higher speeds, but since the 1950s, with the introduction of new wing designs that lower resistance and jet engines for propulsion, it was possible to travel faster than sound.

When an airplane flies, it pushes air molecules and drives them out of its way, continuously creating waves of compressed and expanded air. These waves are sound waves, which move away from the plane in all directions at about 1,125 feet per second, the typical speed of sound. If the plane travels at a slower speed, sound waves can propagate in front of it, but if the plane reaches the speed of sound, the waves begin to accumulate in the front of the plane and compress, forming shock waves. At the exact moment when the plane reaches the barrier of sound, the waves that have been accumulated and compressed at the front burst into what is known as a supersonic explosion.

Sometimes when an aircraft breaks the sound barrier, a cloud is formed, caused by a fall in pressure. This drop in pressure implies a drop in temperature. If the air is humid, water vapor condenses in small droplets and forms the cloud.

In aerodynamics, it was believed that the sound barrier was a physical limit that prevented large objects from moving at supersonic speed.

ART

used
+

COPY

THE INFINITE WORKS OF M.C. ESCHER

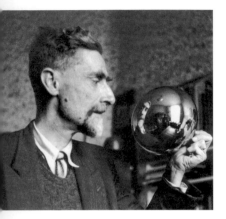

Maurits Cornelis Escher (1898–1972) was a Dutch artist known for his wood engravings, woodcuts and lithographs of impossible figures, tessellations and imaginary worlds. Escher experimented with various methods of capturing paradoxical spaces that challenge habitual modes of representation in drawings of two or three dimensions, which has meant he has become the favorite artist of many mathematicians. The numerous reproductions of his works in books, magazines and advertising campaigns have made Escher one of the most referenced artists in popular culture of the 20th century.

As an artist, M.C. Escher is difficult to classify. There have been many interpretations of his works, but Escher had no major pretensions or messages to transmit, he basically expressed what he liked. His work was not based on feelings, like other artists, but only on situations, solutions to problems, visual games and nods to the viewer; visions sometimes struck him at night, passing through his imagination and which he thought worthy of being reflected in his paintings. Though he was not a professional mathematician, his works show an interest and a deep understanding of geometric concepts, from the perspective of curved spaces, via the division of a plane by repeated symbols.

His works also feature the impossible figure known as the tribar, a three-dimensional triangle, which is actually impossible to build; it only exists as a drawing. Escher learned about this design and worked from it. He illustrated designs of several buildings that implicitly used one or more tribars and for which, after careful observation, we can prove the impossibility of the drawing, like stairs that come from the inside of the building but that lean on the façade in *Belvedere*, or a water wheel in perpetual motion in *Waterfall*. In the latter, if you look at the stream of water, you can continuously follow it without ever reaching the end, representing an infinite loop.

Another impossible design that Escher used to demonstrate continuous movement is the Penrose stairs. With it he created his famous works featuring stairs that further explore the idea of endless movement. These optical effects are due to the human mind's misleading tendency to perceive in three dimensions while we are observing a two-dimensional image.

Another curiosity by M.C. Escher is his self-portrait Hand with Reflecting Sphere: *what the artist actually used was not a sphere but a bottle with a spherical bottom.*

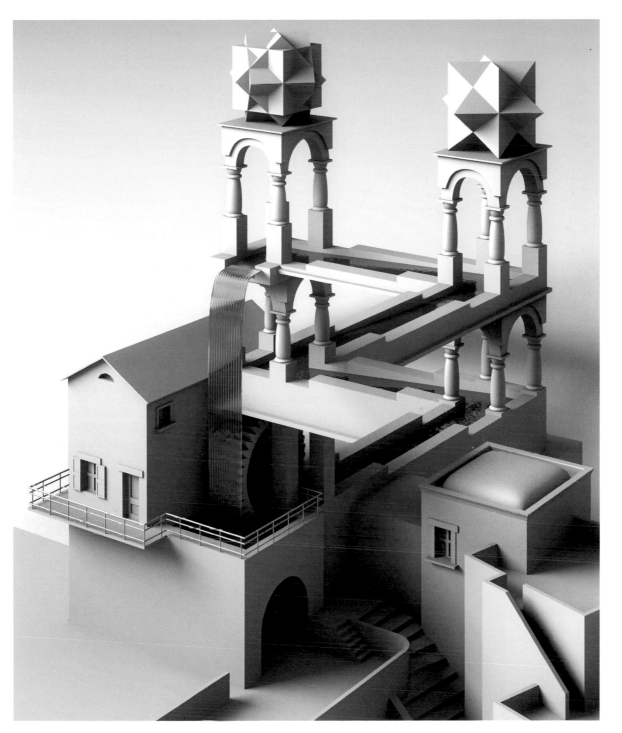

THE ENDLESS STAIRS OF THE VATICAN

The Spiral Staircase of the Vatican Museum is one of the most photographed in the world, and certainly one of the most beautiful. Its helical construction is somewhere between a ramp and a staircase, designed so that horses could go up and down in situations of emergency, as it was used years ago as both an entrance and exit. The constant crowds of people meant the staircases were often full, creating a trompe l'oeil effect: it was not known why the people going up the stairs and those going down never crossed paths until the construction was analyzed.

The staircase as it exists today was debuted to the public on December 7, 1932. Momo is the architect of the current staircase, though its construction was based upon the original design by Bramante, and the balustrade is credited to Maraini. The engineer and architect Giuseppe Momo (1875–1940) built several buildings in Turin, Piedmont, but above all in Rome, where Pius XI commissioned him with the architectural transformation of the Vatican City. He completed more than 200 architectural projects, including the Governor's Palace of the Vatican, the train station, the headquarters of the Pontifical Lateran University and the construction of the entrance to the Vatican Museums, which is a double spiral staircase and a glass ceiling.

Until recently no attention had been paid to the fascinating optical effect caused by walking up and down the staircase, as it actually seems endless. The logical explanation for this is that it produces an effect that is achieved by means of a double helix coiled to the right from top to bottom, as it is not one but two twisted staircases, one to go up and one to go down, creating a structure similar to DNA and causing the sensation of eternal stairs.

Donato Bramante (1444–1514), the architect who around 1505 designed the ramp which Momo used as a basis to develop the ensemble, is considered the pioneer of the Roman Renaissance, a style characterized by sobriety, classic beauty and clarity.

WAGNER'S INFINITE MELODY

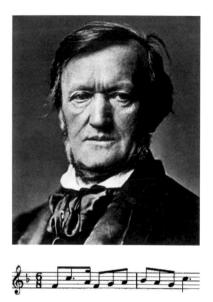

Wilhelm Richard Wagner (1813–1883) was probably the most important musician of the second half of the 19th century. He is known for his operas, described as "musical dramas" by the musician himself; and unlike most composers, he also wrote the lyrics and managed the stage design. His aesthetic design, based on the concept of *Gesamtkunstwerk*, "total artwork," had a great impact on all contemporary music and his influence can be seen in philosophy, literature, visual arts and theater.

According to Wagner's concept of total artwork, the musical drama should combine the elements of the word, music, and action into a single work unit, a symphonic flow uninterrupted by a process of change and development. For Wagner, the focus is on action, which is presented through music and words so that none of these three elements can be separated from the others. Thus neither word nor music can take supremacy in musical drama. Both elements must collaborate and interact to manifest and show the viewer the action, which represents the true realization of artwork. For this reason, Wagner believed that the music and the lyrics must be written by the same person.

For the work to have maximum intensity, both verbal and dramatic, it was necessary for its musical structure to retain a character of total continuity: it should be *durchkomponiert*, that is, composed with an uninterrupted flow, without returning or repeating. One key device used is alliteration, the repetition of heavily accented consonants at the start of two or more successive words. In Wagner's alliterative verse, the initial number of accents in each line was irregular, causing the end of the regular quadrature of the musical period (phrases of 4 or 8 bars, common in the late 18th century); in other cases, irregular melodic phrases were alternated between voice and orchestra, linking one to the other almost without interruption, resulting in what is known as infinite melody.

The infinite melody was able to inspire the brightest minds, most notably the German philosopher Friedrich Nietzsche, who maintained a love-hate relationship with Wagner for decades.

ANDREA POZZO AND THE INFINITE

The 17th century represents a break with the values of universal order that had been the basis of scientific thinking. With the rise of scientists such as Kepler, experimentation, hypothesis and scientific reasoning gained supremacy over more traditional notions of the universe. Infinity appeared as pictorial space. So it is no wonder that the illusory, already used from the late 15th century, was systematized.

The Jesuit Andrea Pozzo (1642–1709) was an Italian painter, architect and writer, famous for his magnificent architectural perspectives, frescoed domes and vaults in churches and palaces using the illusionist technique of *quadrature*, in which he combines architecture, sculpture and painting. His art is summarized in the treatise *Perspectiva pictorum et architectorum*, a publication that would become one of the fundamental books of the Settecento scenery.

Pozzo's artistic activity focused on the decoration of several new Jesuit churches, which were experiencing a boom period. So, Pozzo adorned many of the churches of the order such as those in Modena, Bologna and Arezzo. In 1676, he decorated the interior of San Francesco Saverio in Mondovi (Piemonte). His more modern illusionist techniques can be seen in this church: the false gold, bronze colored statues, marbled columns and trompe l'oeil dome made on a flat roof, decorated with figures in architectural settings.

His most outstanding work is the ceiling of the vault of the Church of St. Ignatius in Rome (1685–1694), which displays a technique that has surprising optical effects. In this vault, the painter shows an impressive mastery of perspective, as he extends the real architecture of the building with a feigned architecture. Also, there is a mark on the floor of the temple where the viewer should stand so they get the most out of the Baroque features.

The idea of perspective was one of the issues that most concerned the Baroque artists. Andrea Pozzo demonstrated a mastery of this technique, with which he managed to create infinite space effects in the frescoes of the vaults of many buildings.

EDGAR ALLAN POE'S EUREKA

Edgar Allan Poe (1809–1849) was an American romantic writer, recognized as one of the universal masters of short stories, of which he was one of the pioneers in his country. He reinvented Gothic fiction, and is especially remembered for his tales of terror. He is considered as the inventor of detective fiction, and he contributed several works to the emerging genre of science fiction.

We all know some of his stories, classics like *The Black Cat*, *The Tell-Tale Heart* or *The Raven*. However, Poe had another side to him unknown to many: from a young age, he was interested in science and devoured any publication by any author on astronomy and cosmology (Newton, Kepler, Laplace, etc.). In fact, he discovered the true explanation of Olbers' paradox and proposed that the origin of the universe was an explosion, now known as the Big Bang.

In 1848, after the death of his wife, Virginia Clemm, he wrote a short essay entitled *Eureka, a prose poem*, which he dedicated to the scientist Alexander von Humboldt. Here, Poe tells us about the cosmos in all its magnitude, metaphysics, astronomy, mathematics, but also the spirit of the universe and how everything has a single purpose: to find the ultimate truth.

With this essay, Poe believed he was contributing to the history of human science. However, this writing was not science based, but rather its subject was the poetry that permeates the universe. Obviously, it was criticized by scientists both contemporary and present-day, as it describes a kind of boundless vision of the future. Branded by many as almost insane, his recovery from this was only possible as a result of the devotion that the French symbolists had for the author, led by Baudelaire.

The master of horror stories and Gothic literature was a big fan of science. The poem Eureka *is a cosmological theory that seems to presage the Big Bang, the theory of relativity, black holes and the first known solution to Olbers' paradox.*

THE METRONOME

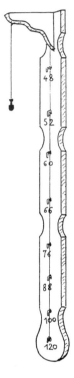

A metronome is a tool that helps the musician to keep a steady tempo in composing musical scores. It produces a sound similar to a ticking clock that sets the tempo indicated for accurate interpretation of the notes that are part of a musical composition. In the classical metronome, speed, measured in beats per minute, can be varied by a movable weight placed on the pendulum of the device, although there are also electronic metronomes, in which the rate is adjusted by means of a dial or a series of buttons. Formerly, metronomes consisted of a pendulum with an adjustable pulley to determine a faster or slower time.

Before this invention, composers used the average human pulse as a reference speed (about 80 beats per minute). This practice dates back to 1812, when it was conceived by Dutchman Dietrich Nikolaus Winkel (1780–1826), who despite being its discoverer, did not have the foresight to register it, so another fellow countryman, Johann Mälzael, copied many of the ideas and was awarded the patent of the portable metronome in 1816. The composer Ludwig van Beethoven (1770–1827) was the first to establish in his musical compositions markings of time by means of a metronome.

It should be noted that many pieces of music have an indication of the tempo to follow at the top of the sheet music. Formerly, to establish the pace of a composition subjective terms such as *allegro, vivace, andante, presto* were used, but today it is customary to note the duration in exact terms, through the use of this device.

Music students often use a metronome to practice, and that helps them adhere to a standard time. A rhythm meter is useful for people with hearing or psychomotor impairments. It is a visual metronome marking time through an animated graphic in which the duration of the beats can be fully understood.

The metronome is used to indicate accurately the tempo (speed) at which a piece of music is to be performed. To vary the speed of the metronome, the top counterweight has to be moved: higher to mark a slower tempo, and lower for a faster tempo.

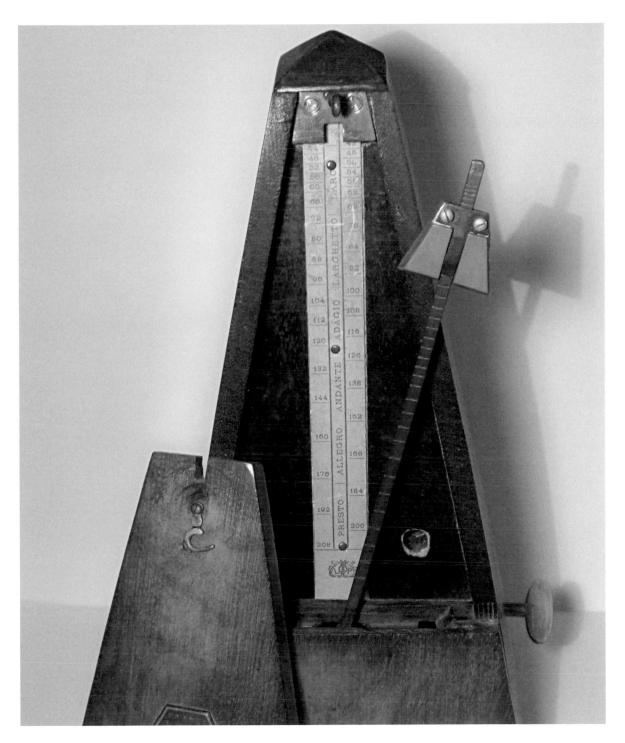

EDVARD MUNCH'S INFINITE SCREAM

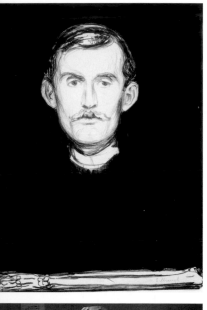

The Scream, painted in 1895, is one of the most famous paintings by Edvard Munch (1863–1944) and indeed of the expressionist movement, and it is one of the most recognizable cultural icons of our time. Its bright colors, emphatic brushstrokes and twisted lines make it one of the more realistic artistic expressions of anguish and pain ever.

This painting reflects his own fears and torture. Its expressive power is due largely to the painting techniques and effects used, the stridency of colors and sinuous lines. Munch carried out several versions of this work. They all show an androgynous figure in the foreground, which is a modern man in a moment of deep anguish and existential despair. The background scenery is Oslo seen from the hill of Ekeberg.

The first version, entitled *Despair*, portrays a man with a hat, but apparently the artist did not feel that the work truly reflected the dark moment that he was going through, so he created a second version with the same name in which he draws a less human figure, facing the viewer and revealing true desperation, rather than mere somber contemplation. It is believed that this expression may have been inspired by a Peruvian mummy that he saw in 1889 at the World Exposition in Paris.

To understand the essence of *The Scream*, one has to go back to Munch's childhood, marked by a strict father and feelings of abandonment and helplessness resulting from the death of his mother and one of his sisters, who was a victim of tuberculosis. Later, his other sister was hospitalized in a psychiatric hospital because of bipolar disorder. The artist lived periods of sanity and madness, aggravated by alcoholism, and his mood is reflected in these lines that Munch wrote in his diary around 1892:

"I was walking along a path with two friends. The sun was setting. Suddenly the sky turned blood red, I paused feeling exhausted, and leaned on the fence. There was blood and tongues of fire above the blue-black fjord and the city. My friends walked on, and I stood there trembling with anxiety and I sensed an infinite scream passing through nature."

Edvard Munch is considered one of the most important painters and writers of the expressionist movement for the sad and distressing representations of their personal obsessions and frustrations.

THE DROSTE EFFECT

An image with Droste effect, also known as *mise en abyme* (equity gap) shows a smaller version of itself within it, which in turn included an even smaller version in a similar place, and so on.

This effect is called Droste after the Dutch company, whose traditional metal container shows a nurse holding a silver tray with a cup of hot chocolate next to a Droste cocoa container. In this package, obviously, you can see the nurse, the silver tray and Droste packaging, in which the image is repeated.

The name comes from French heraldry, where often, the shield includes in its center, the *abyme*, an image of itself in miniature. Naturally, if we continue, we will be trapped in an endless chain, since within each shield there is always a smaller image in the center of itself, with another smaller shield in the center.

This repetitive motif can be continued in reduced size only in theory, as in practice it is limited by the resolution that the printing technique used is capable of, since each iteration exponentially reduces the size of the image.

It is rare to find this visual effect in advertising or magazine covers, and the Droste effect is not a recent idea. It was used, for example, by Giotto di Bondone in 1320 in *Stefaneschi Triptych*, which is now in the Vatican Museums. There are also some books from the Middle Ages that recursively repeat its own image, and there are examples of stained glass showing miniature copies of the same stained glass.

One way to generate a version of the Droste effect is to place two mirrors so that they mirror each other. If someone passes in front of the mirrors, they will be reflected infinitely.

The term "Droste effect" was coined by the poet and columnist Nico Scheepmaker at the end of the 1970s although in the 1950s the artist M.C. Escher had already popularized it in his works.

SKIP

LEOPARDI'S INFINITE

Count Giacomo Leopardi (1798–1837) is one of the most important Italian poets from the Romantic era and one of the leading figures in literature. His work is characterized by philosophical meditation and introspection.

One of his early poems *The Infinite*, composed in September 1819, deals with one of Leopardi's fundamental issues. The poet used to go up to the Recanati hillsides, in his home city, to ponder this landscape from which he drew a lot of inspiration. A hedgerow blocked his view of the horizon and this obstacle allowed him to fantasize. Beyond the hedgerow he imagined a space with no limits, profound silences and absolute peace, so much so that he felt dismayed. Then the sudden sound of the wind among the plants brought him back to reality.

Placing poetry in its cultural context, one can understand Leopardi's thinking and his poetic concept of infinity. Man, by nature, is made for the infinite and yet he does not know how to reach it. What man wants is happiness, that infinite pleasure which, nevertheless, he cannot achieve, given that everything that surrounds him is finite and limited. This results in the continuous feeling of dissatisfaction and desire to surpass these limits. Tedium is the "sublime" feeling of dissatisfaction where nothing on earth is enough for man and derives directly from the consideration of the infinite.

This contradiction between infinite desire and the inability to achieve sustained happiness is the basis of all the work and thinking of Leopardi: man questions the meaning of life but cannot find answers if he is not accepting of illusions, even though they are false. The condition of man is, therefore, unhappiness, despair. However, the poet always makes his desire not to surrender clear, to continue to seek the happiness that he knows is unattainable.

Always to me beloved was this lonely hillside
And the hedgerow creeping over and always hiding
The distances, the horizon's furthest reaches.
But as I sit and gaze, there is an endless
Space still beyond, there is a more than mortal
Silence spread out to the last depth of peace,
Which in my thought I shape until my heart
Scarcely can hide a fear. And as the wind
Comes through the copses sighing to my ears,
The infinite silence and the passing voice
I must compare: remembering the seasons,
Quiet in dead eternity, and the present,
Living and sounding still. And into this
Immensity my thought sinks ever drowning,
And it is sweet to shipwreck in such a sea.

CARDINAL POINTS AND THE INFINITE DIRECTIONS

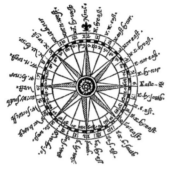

To orient or locate a place cardinal points are used. They have a direct relationship to the Sun's apparent motion in the sky throughout the day, following the rotation of the Earth. They also comprise the Cartesian reference system universally used to represent the orientation in the various plans and maps of Earth's surface. The point and straight line has always been used to describe a position in space, for a predetermined coordinate system. From any point within a coordinate axis, there are an infinite number of straight lines with infinite paths.

Different cultures have different values and symbols assigned to each of the directions represented by the four cardinal points, determined by the position of the northern pole, by the position of the Sun at noon, the south and by the rising and setting of the Sun at the equinoxes, the east and west respectively. These four directions are four 90-degree angles, which in turn are subdivided with bisectors, generating the northwest, southwest, northeast and southeast, and by repeating the same operation, the popular wind rose is obtained, used in navigation since ancient times and that covers the main 32 directions of the Earth's surface as well as the direction of the wind.

The word cardinal comes from the Latin name *cardo*, which identified a street laid out from north to south in Roman cities. This means that the only truly pivotal point, at least from the etymological point of view, should be the North and to a lesser degree the South.

In Norse mythology, Norðri, Suðri, Austri and Vestri were four dwarves who each support one of the four corners of the world and their names are derived from the four main cardinal points (north, south, east and west, respectively).

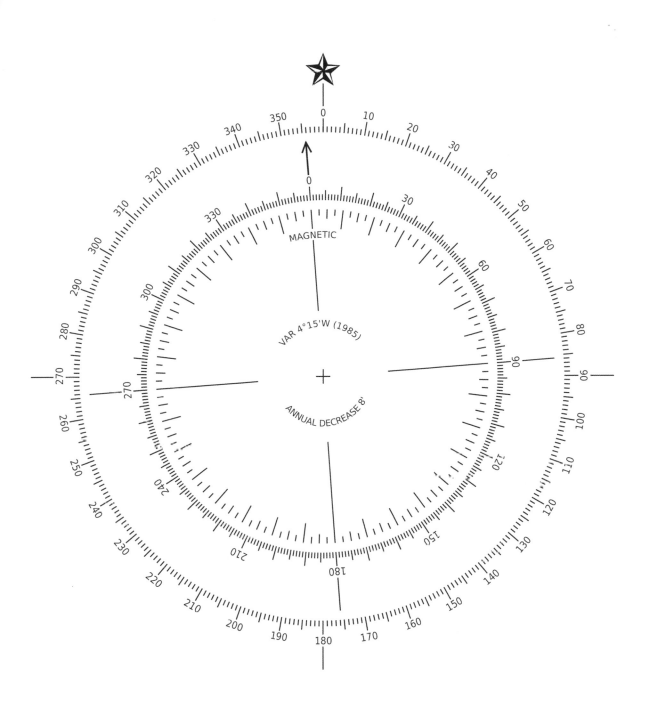

PARALLELS AND PROJECTIVE GEOMETRY

When we say that two parallel lines intersect at a point, we are building on the principles of projective geometry, a mathematical model to study the intuitive concepts of perspective and horizon regardless of the measurement concept. It is based on two clear premises: two points define a line and any two lines intersect at one point. This cutoff point is called the ideal point or point at infinity.

The early stages of the graphical representation in three dimensions appear in the golden age of human knowledge, the Renaissance. Since antiquity, man has felt the need to graphically capture the surrounding environment, as evidenced by cave drawings, but until the Renaissance, with artists such as Leonardo da Vinci, Albrecht Durer or Piero della Francesca, preceded by Giotto, no attempts were made to represent depth. So, the need emerged to establish the formal basis on which to lay the new forms of geometry to accommodate depth: projective geometry. Its fundamental principles were established by the mathematician Girard Desargues (1591–1661), whose work went unnoticed for two centuries, probably overshadowed by the genius Descartes.

Projective geometry includes as a fundamental property the incidence between any two lines: in the projective plane, two distinct lines *L* and *M* always intersect at exactly one point *P*. Contrary to what happens in traditional geometry, in projective geometry, parallel lines do not exist. The classic postulate (Euclidean) of parallel lines is removed by adding the point at infinity or ideal point to the plane. Thus, two parallel lines have an ideal point in common, which can be thought of as its direction. These new points form a line in turn, called the line at infinity or ideal line, or, horizon.

Girard Desargues was an architect and military engineer passionate about mathematics applied to architecture and painting. He was used to writing his ideas on loose-leaf pages which he distributed among his colleagues and that he did not publish. For this reason a lot of material was lost.

use

COPY

THE PAINTER OF NUMBERS

The passage of time became an obsession for the French-born Polish artist Roman Opalka (1931–2011). The artist conceived time as an irreversible continuum flowing through the person, the pulse of life approaching death through the greatness of the infinite. So, one day in 1965 he began to inscribe a succession of numbers in his studio in Warsaw, painting with a trembling hand the number one in the upper left corner of the completely black canvas, and that would extend to infinity or to where he was destined to reach. When he died he had reached 5607249. *5 million 6 hundred seven thousand*

Opalka painted a total of 233 illustrations, which form a series dubbed *OPALKA 1965 / 1-∞*. In the 46 years of his Herculean task, which for some critics was nothing short of a prolonged suicide, he used, invariably, canvases of 196 by 135 centimeters, and the numbers drawn were always identical, drawn with a number zero paintbrush. Each figure was only a centimeter high, each painting continued the previous exactly where he left off. In 1968, he went from a black to gray background, and in 1972, when he reached a total of 1,000,000, he gradually began to lighten it by adding 1 percent more white each year. In 2008, he ended up painting white figures on a white background, a shade he called *blanc merité*, white well earned.

Also in 1972 he began to record himself with a tape saying he was painting the numbers, about 380 a day and between 20,000 to 30,000 per canvas. When he finished painting each canvas, he took a photo in front of his recent work, always in the same conditions and lighting techniques to record the parallelism between the increasing sequence of numbers and the aging of the artist as he approached infinity.

It is said that Opalka's obsession started in 1965 when one afternoon he was waiting for his wife at a cafe in Warsaw and she was late. It occurred to him that he could express time through painting.

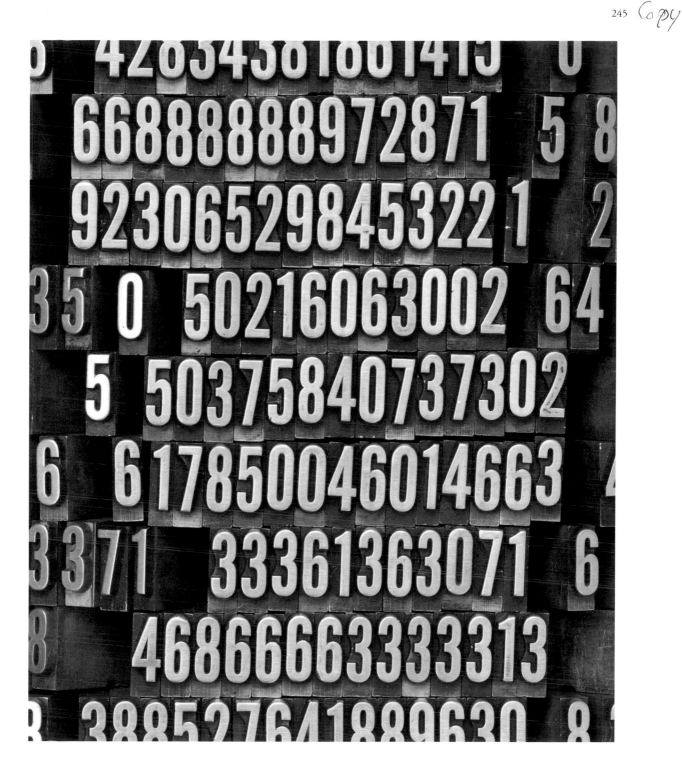

use

THE INFINITY OF JORGE LUIS BORGES

As we know, the representation of infinity has concerned mankind for centuries in various fields of artistic expression. In the case of literature, Jorge Luis Borges (1899–1986) stands out, arguably the greatest Argentine writer with universal relevance, who used this concept as a theme in several of his stories.

Borges chose the short story format to convey his idea of infinity because to him the infinite is not only unattainable, but it is also inconceivable. According to Borges, infinity is a negative concept, and he expresses this view in several stories through the theme of infinity as the realm of chaos and uncertainty, such as in *The Library of Babel* and *The Book of Sand*. This unexpected notion differs from those that mathematics presented, in which infinity is associated with a progression of unreachable numbers. According to the Argentine writer, infinity was something else.

In *The Book of Sand,* he presents a view of the infinite that defies any regulation and that is above any order and all predictions. This metaphor of infinity takes the form of a mysterious book that the main character receives from the hands of a stranger. This is a book that, like sand, has neither beginning nor end: the number of pages in the book is exactly infinite. The numbering of the pages is not correlative, and once you turn a page it is impossible to find it again. The protagonist describes his vain attempt to find the first page. This book eventually becomes the obsession of the character.

A huge fan of metaphysics, the writer was interested in the infinite theory by the mathematician Georg Cantor who introduced transfinite numbers, represented by the Hebrew letter aleph, which also served as inspiration for Borges for the title of a story.

∞

NICCOLÒ PAGANINI AND *PERPETUUM MOBILE*

The virtuoso and composer Niccolò Paganini (1782–1840) is recognized as one of the greatest violinists who ever lived. He applied his own methods to the interpretation of the violin and developed a technique superior to that typical of his time, to the point where it was said that he had supernatural powers through a pact with the devil. He could interpret works of great difficulty with only one of the four strings of the violin (first removing the other three), and continue so that it seemed as though several violins were being played.

He performed throughout Europe and achieved success with his own pieces and variations on opera. He composed concertos for violin and orchestra, some only in an outline form, and besides the two complete concertos, another four have been reconstructed. He also left 24 capriccios for violin and several sonatas.

It is said that Paganini performed one of his works in 3 minutes, *Moto Perpetuo*, a composition that to this day nobody has been able to play in less than 4½. The musical term that gives name to this work, known as perpetual motion, or *perpetuum mobile* defines a type of composition where the entire piece, or part, is repeated a number of times, often undefined, without the movement of the melody stopping when a repetition begins.

The *perpetuum mobile* as a musical genre reached its height of popularity in the late 19th century. The pieces are often performed in the encore of concerts, in some cases increasing the tempo with repetitions.

His skill with the violin was so magnificent that it awoke rumors of all sorts which he never wanted to deny, from a pact with the devil to diseases that gave him unusual flexibility in his hands.

THE MODERN ORPHEUS,

Opera House   June 3rd 1831.

Sketches of the Musical World Nº 1, to be continued.

Published by Thoˢ. Mᶜ Lean, 26, Haymarket, June 10th 1831.

SOLOMON'S KNOT

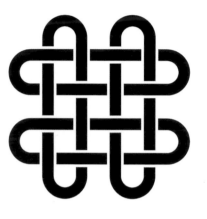

According to biblical tradition, Solomon was the wisest of the kings of Israel; he received from God the ability to distinguish between good and evil, and built a temple that was the representation on Earth of the union between the divine and the human. In the early Middle Ages and following centuries, any object or symbol attributed to Solomon was a sign of prestige entailing something magical, powerful and highly positive.

Along with other symbols such as the circle or spiral, Solomon's knot can be identified as one of the oldest and most widespread: it is present in ancient rock art, in regions distant from each other and in all the great cultures from Roman to Germanic, from Jewish to Islamic, from Indian to various African cultures. It is very common in early Christian art, and several examples of Solomon's knot can be found among fragments and remnants of the early Roman basilicas. Since reaching its peak in the Celtic culture, based on the themes of knots, weaves and undulated shapes, it is thought that it was the Romans who extended the use of the knot through their dominions soon after their contact with the Celtic culture.

Solomon's knot consists of two elements, usually two joint, elongated rings, sometimes in the shape of a shield, rectangle or pointed arch in the Middle Ages, representing the interconnection and the union of two elements. Originally, it represented the union between the divine and the human, an unmistakable sign of the old secretive adage "the bottom is like the top and the top is like the bottom." Its main feature is the absolute symmetry between top, bottom, left and right, because regardless of the side from which it is observed, it maintains the same form and meaning.

Along with the spiral and the circle, Solomon's knot is one of the oldest symbols in existence. It has been used for centuries in handicrafts, architecture and other works of art in general.

ART

YAYOI KUSAMA AND HER INFINITY NETS

Yayoi Kusama, born in 1929, is a Japanese contemporary artist who throughout her career has worked with painting, collage, sculpture, performance and facilities, showing her thematic interest in psychedelic colors, repetition and patterns. Although her work is varied and irregular, Kusama's best works, which form the series *Infinity Net*, are among the most complex paintings at a visual level and most provocative at a conceptual level.

The images and hallucinations that had tormented her since childhood became obsessions that took shape in her works. The most overwhelming was probably an infinite profusion of polka dots covering the surface of all things. Maybe that's why her *Infinity Nets*, large expanses of repetitive polka dots on paintings, objects and environments, often extended with mirrors, are both so fascinating and disturbing.

In 1957, she moved to New York and produced the first painting of *Infinity Nets*, upon a large blank monochrome canvas, initiating a series of paintings of different sizes, which at times reached a height of 36 feet (11 m), made with a simple flick of the wrist that is repeated endlessly as a network of polka dots and which are the expression of her obsession with infinity. In 1960 and 1961, she continued in the same vein, but introducing color. The artist describes *Infinity Nets* as paintings "without beginning, end or center. The entire canvas is covered with a monochrome network. This infinite repetition causes a hypnotic sensation of vertigo, of emptiness."

In 1973, Kusama returned to Japan and disappeared from the art scene as suddenly as she appeared onto it. In 1977, she was diagnosed with obsessive neurosis and voluntarily entered a psychiatric facility, where she still resides.

Other famous series by Yayoi Kusama are the Accumulation Sculptures *and* Infinity Rooms*, in which the accumulation and repetition of forms are developed. Polka-dots and spot colors are key elements.*

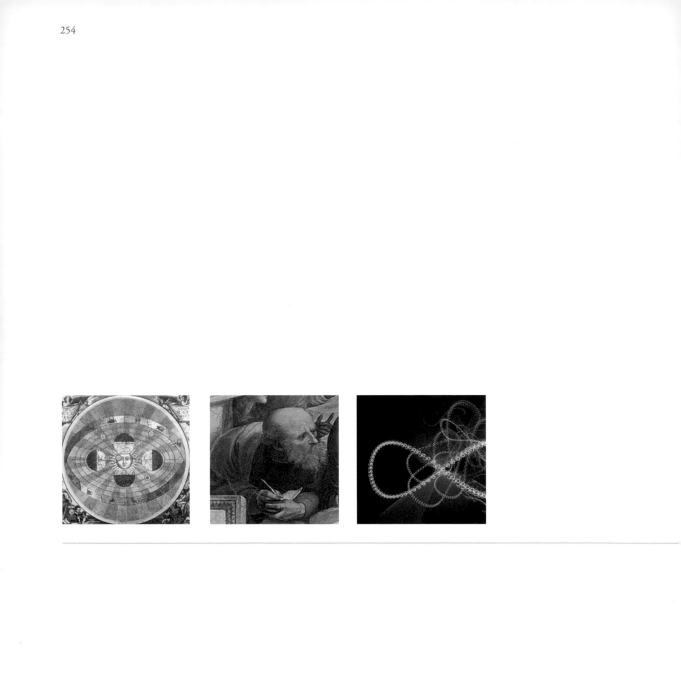

PHILOSOPHY

SKIP

ANAXIMANDER AND THE INFINITE COSMOLOGY

Anaximander of Miletus (610 BC–546 BC) was an Ionian philosopher. Disciple and successor of Thales of Miletus, he is credited with only one surviving book about nature, but his word has reached us through the doxographical comments from other authors. He is also the author of a map of the land, in addition to measuring the solstices and equinoxes by a *gnomon*, work to determine the distance and size of the stars, and originator of the claim that the Earth is cylindrical and occupies the center of the universe.

The answer given by Anaximander to the question of *arche* certainly exceeds the hypothesis of Thales of Miletus. The *arche*, the primal element, is now the *apeiron* (*a*, primitive particle, and *pears*, limit, perimeter), that is, the indeterminate, the unlimited, the infinite. He claims that this is not water or any other of the so-called elements, but some other *apeiron* nature without limits, without definition, immortal, indestructible, imperishable and unengendered from which all the heavens and the worlds in them generate. However, where there is generation, there is also destruction.

For Anaximander, cosmology describes the formation of the universe through a rotation process that separates the hot from cold. The fire occupies the periphery of the world and can be seen through the holes that we call stars. The Earth, cold and wet, is at the center. The first animals emerged from the water or mud heated by the Sun, later moving to Earth, and people descended from fish, an idea that certainly is an astute anticipation of the modern theory of evolution.

Anaximander of Miletus is credited with the first map and the invention of cartography. Although unfortunately it has not survived to this day, we know that the map included the whole habitable earth with all the known seas and rivers.

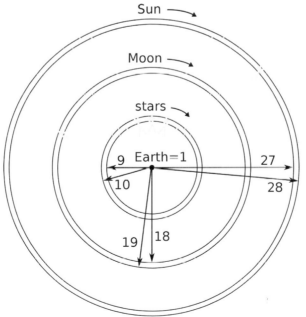

COPERNICUS AND THE HELIOCENTRIC VIEW

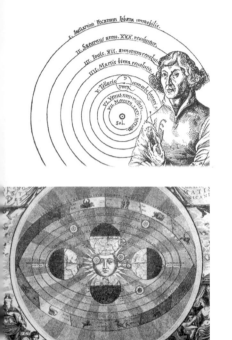

Nicolaus Copernicus (1473–1543) formulated the first heliocentric theory of the solar system and his ideas marked the beginning of what is known as the scientific revolution. He is considered the founder of modern astronomy thanks to his book *De revolutionibus orbium coelestium* (*On the Revolution of the Celestial Spheres*), a study to which he devoted more than 20 years.

This work describes the Copernican system, or heliocentric theory of the universe, with the mathematical proof necessary for its justification. The radical astronomical change proposed by the Copernican theory consists of placing the sun at the center of the universe, and making the spheres of Mercury, Venus, Earth, Moon, Mars, Jupiter and Saturn spin in its orbit. Earth is now considered just another planet, the third in distance from the Sun, rather than the immobile center of the universe, as the whole Aristotelian and Ptolemaic cosmology (except Aristarchus of Samos), had assumed. Copernicus asserted that our planet has three movements: daily rotation, annual revolution, and annual tilt of its axis. According to this model, the distance from Earth to the Sun is small compared to the distance to the stars, which remain immobile; they do not revolve around the sun.

This model is one of the most important theories in the history of Western science. The change, which represented for medieval religious ideology the replacement of a closed and hierarchical cosmos, with man as the center of the universe, with an infinite and homogeneous universe located oriented the Sun, made Copernicus doubt on whether or not he should publish his work, conscious of the problems that a disclosure of such magnitude could lead him to have with the Church. Because of an illness that caused his death, he failed to see it published.

The first edition of *De revolutionibus* appeared in 1543, which introduced the work dedicated to Pope Paul III, justifying the authorship of the work with the clear failure of astronomers of the time to resolve and agree on a theory on the rotation of the planets, with emphasis on the accuracy of their findings and predictions, which would allow the Church to solve one of the most important problems. This problem was none other than the development of a more accurate calendar, a fundamental task which served to encourage sponsorship and funding of astronomy by the Church.

Despite the initial rejection, the Copernican model was widely accepted from the 17th century up until the 20th century, when it was discovered that neither the Sun nor the Milky Way are at the center of the universe.

SKIP

GIORDANO BRUNO AND THE INFINITE

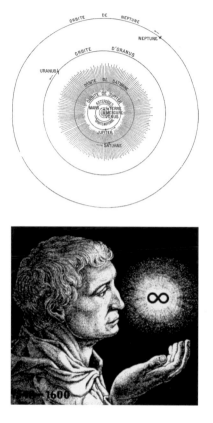

Giordano Bruno, born Filippo Bruno (1548–1600), was a religious philosopher, astronomer and poet who after studying in Naples majored in humanities and dialectic. In his works, there is an obvious influence of Nicholas of Cusa, but also of Plato and the Neoplatonists and even Pre-Socratic influences. Although he was interested in the art of memory and logic, he was primarily a philosopher of nature.

It is known that he had read the work of Copernicus *De revolucionibus orbium coelestium* (1543), almost unknown at that time. In his opinion, Copernicus did not delve far enough into clarifying the full implications of the heliocentric theory, since it is a mathematical reading that does not go into enough detail to discover the full metaphysical implications.

Bruno brought the heliocentrism of Copernicus to its most extreme consequences: the whole Aristotelian system was false, in other words, the sky does not exist, the universe is infinite and there are infinite worlds. A rupture was produced in terms of the Greek conception in which the perfect is finite and limited; also "universe" and "world" are no longer synonyms to include one in the other. It is thus impossible to determine what is the center of the universe and even more difficult to ascertain its circumference. There are no transparent spheres: the stars drift freely through space and the heavenly regions disappear because all the stars are composed of the same elements.

To support his thesis, Bruno advanced the premise that a finite universe does not correspond to the infinite power of God, since it makes no sense that God had limited his creative power. Beyond that, Bruno argued that the world moves spontaneously, instead of using the Aristotelian driving forces. The universe is thus like a giant animal, in the sense that everything is animated.

Bruno's ideas regarding the infinity of space, movement of the stars, the plurality of worlds and heliocentrism resulted in his being persecuted by the Catholic Church and the Inquisition. He was jailed in 1592 on charges of blasphemy, heresy and immorality, condemned as heretical, unrepentant, stubborn and opinionated and burned at the stake in 1600 in Rome. No doubt his death had a discouraging effect on the scientific progress of civilization, but nevertheless his scientific observations continued to influence other thinkers and he is now considered one of the pioneers of the scientific revolution.

More than 400 years ago, modern science was anticipated by his theories of an infinite universe with infinite worlds.

Use?

COPY

KANT AND THE ANTINOMY OF INFINITY

Immanuel Kant (1724–1804) was a philosopher of the Enlightenment. He is the chief representative of German idealism and is considered one of the most influential thinkers of modern European philosophy. Among his most important works is the *Critique of Pure Reason*, described generally as a milestone in the history of philosophy and the beginning of modern philosophy.

The four antinomies emerged from his *Critique of Pure Reason*, a kind of paradox that involves two contradictory statements. Immanuel Kant believed that when our capacity for reason goes beyond possible experience, often it falls into several contradictions. Whether the propositions are affirmative (thesis) or negative (antithesis), both can be defended from the point of view of pure reason, and also our experiences cannot confirm or refute neither one nor the other. This is what happens when the ratio crosses the boundaries of experience. Furthermore, Kant notes that the statements expressed in an *a priori* thesis are those of rationalism, while those shown by the four antitheses are typical of empiricism, that is, *a posteriori*.

The first Kantian antinomy on the infinity of the universe is well-known, a contradiction that supports the thesis that the universe has a beginning in time and is limited in space, whereas the antithesis states that the universe has no beginning and no limits because it is infinite in both time and space. Both assumptions could be proven.

The universe must have a beginning because if it did not it would not exist, since everything that exists has a beginning and an end, and it cannot be infinite in space, because as it is something in space, it has to cease to exist and have a limit. But if the universe had a beginning in time and space, what was there before it existed? Because nothing comes from nothing and something had to exist in time, and if it existed before we called it the universe that something should also be included in it.

The error is that space and time have been understood as things in themselves, rather than taking them as ways that our faculty of knowledge applies to the phenomena. The solution of the first antinomy is that both propositions are false, because it uses a theory contrary to the laws and conditions of knowledge as a starting point.

In addition to the Critique of Pure Reason, *published in 1781, the* Critique of Practical Reason *and* Critique of the Power of Judgment *are major works by Kant.*

 COPY

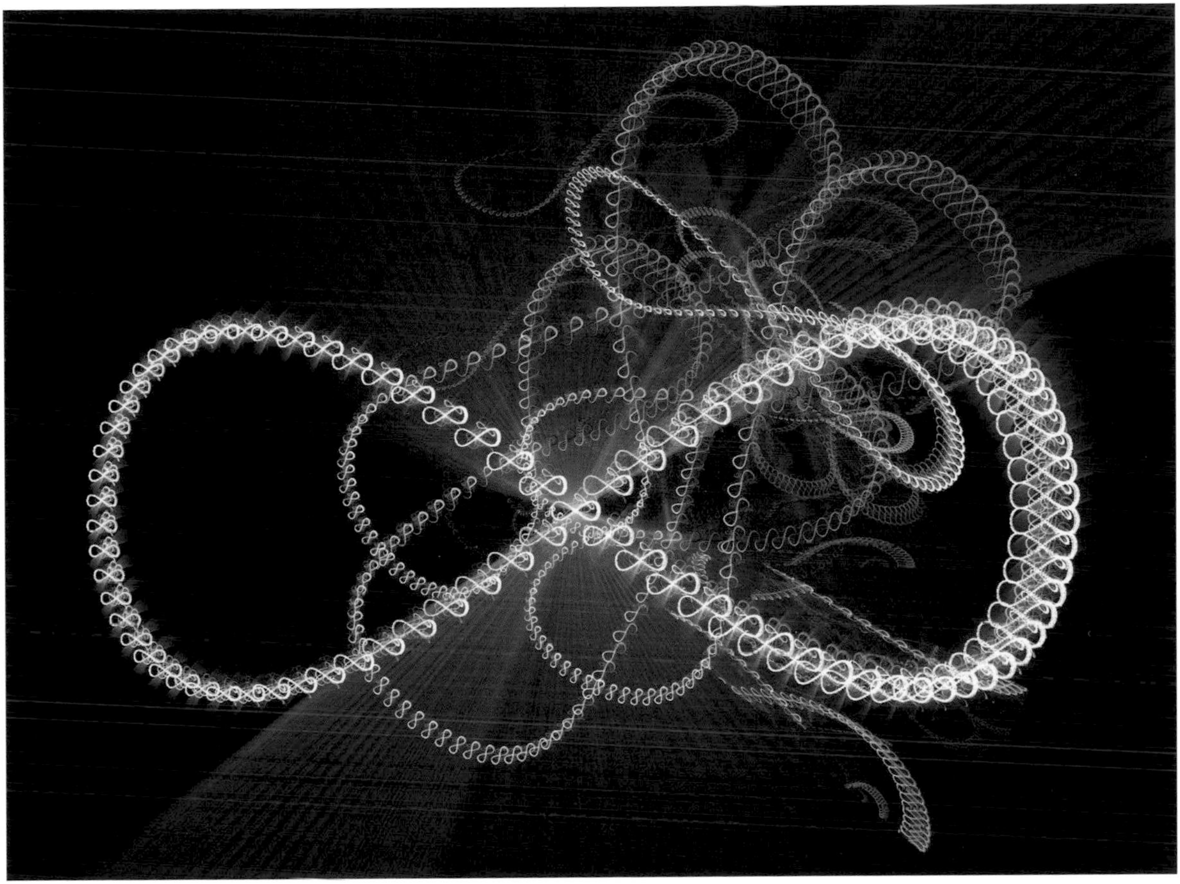

THE HEGELIAN DIALECTIC AND THE INFINITE

Friedrich Hegel (1770–1831) was one of the most important German philosophers of the 19th century and is considered one of the great metaphysicists and the father of German idealism. He received his seminary training at the University of Tübingen, where he befriended the future philosopher Friedrich Schelling and the poet Friedrich Hölderlin. He was fascinated by the works of Plato, Aristotle, Descartes, Spinoza, Kant and Rousseau and by the French Revolution, which he ended up disapproving of when it fell into the hands of the Jacobin terror. Hegel's most important work is *Phenomenology of Spirit* (1807).

For Hegel, the infinite as a concept can primarily be considered a definition of the absolute. Hegel developed his famous dialectic of infinity, his conception of the infinite truth. Infinity should not be conceived as if it were the progression of the finite that, as it progresses, steadily increases its limits: that idea of the infinite is rife with fallacy. The infinite must be conceived dialectically as taking place in the finite and through the finite, where imposed limits are expressed and then denied: the negation of the negation is his claim. According to Hegel, the true infinite is the totality of the moments of being that are determined in each of the limits set by the universal.

For him, the infinite is not beyond the finite, nor is it something empty and indeterminate: the infinite contains, within itself, the finite. The infinite is not transcendent, but immanent in the finite. Therefore, the individual beings, the finite, are merely *moments* of the infinite. The infinite is, therefore, the all or the totality of reality.

From this, it follows that only the real is the all. Something is true only insofar as it is integrated into the totality. The finite, as such, is not true, but rather ideal, that is, something abstract. For Hegel, *concrete* has the etymological sense of a totality that grows and develops, containing within itself its parts, differences and determinations. The *abstract* is a part or moment separate from the all. Only the concrete totality is the truth.

Originally, dialectics was known as a method of conversation or analogous argument which today we call logic. In the 18th century, the term came to mean disagreement over things or concepts as well as detecting and overcoming these disagreements.

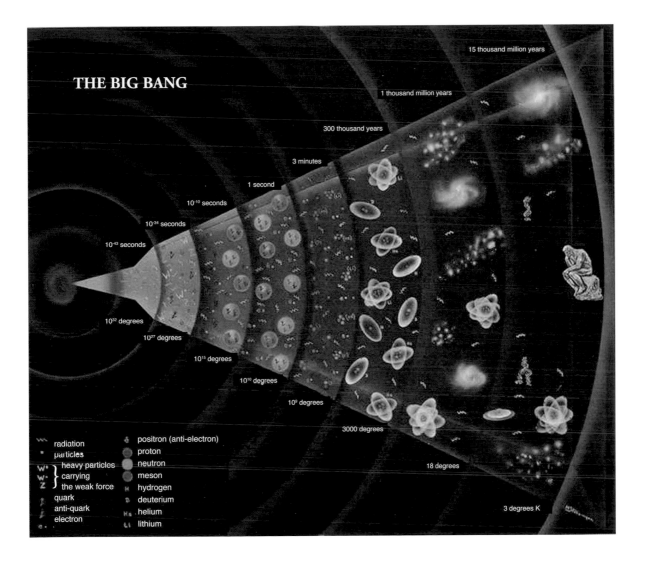

THE BIG BANG

15 thousand million years

1 thousand million years

300 thousand years

3 minutes

1 second

10^{-10} seconds

10^{-34} seconds

10^{-43} seconds

10^{32} degrees

10^{27} degrees

10^{15} degrees

10^{10} degrees

10^{9} degrees

3000 degrees

18 degrees

3 degrees K

ᨠᨠᨠ radiation
▪ particles
W⁺ ⎫ heavy particles
W⁻ ⎬ carrying
Z ⎭ the weak force
quark
anti-quark
e ▪ electron

positron (anti-electron)
proton
neutron
meson
н hydrogen
deuterium
нₑ helium
ʟᵢ lithium

use

copy

PASCAL AND THE TWO INFINITES

Blaise Pascal (1623–1662) was a French mathematician, physicist and philosopher, considered the father of computers along with Charles Babbage. His early works include natural and applied sciences, in which he made important contributions to the invention of mechanical calculators, considerations of the mathematical theory of probability, fluid research and clarification of concepts such as pressure and vacuum by expanding the work of Evangelista Torricelli.

In 1654, he abandoned mathematics and physics to engage in religious and philosophical reflections. The writings and notes from this period were collected posthumously in the *Pensées*, a defense of Christianity published in 1670. Although it is believed that before his death Pascal had planned the structure of the book, in fact he left incomplete work and it is not known exactly which order he intended.

His first concept of infinity is centered upon the universe, which he considered as "an infinite sphere whose center is everywhere and circumference nowhere." Pascal invited readers to contemplate the universe in its vastness to become aware of our limited human condition and of ourselves as individuals.

But he conceived of another infinite, the infinitely small, another abyss that overwhelms us because it is also beyond our powers of understanding, and equally immeasurable. The individual is caught between the infinitely large and infinitely small, both incomprehensible.

Before devoting himself fully to philosophy, Blaise Pascal invented one of the first mechanical calculators, the Pascaline, which operated on the basis of wheels and gears and could perform addition and subtraction operations.

Use

COPY

RENÉ DESCARTES, THE INFINITE AND GOD

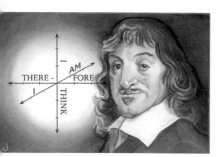

Fig. 1

René Descartes (1596–1650) was a great French philosopher, mathematician and physicist considered to be the father of modern philosophy and one of the most prominent names of the scientific revolution. He used the mathematical method in an attempt to end the Aristotelian syllogism used throughout the Middle Ages as a model.

Descartes proposed the existence of three substances. Our thinking substance (*res cogitans*) is the first and represents the first truth or certainty, the famous *cogito ergo sum*, "I think, therefore I am," an essential element of Western rationalism. Universal and methodical doubt leads the subject to knowledge of the existence of this reality. The fundamental attribute of this substance is thought or consciousness.

The second substance is the infinite (*infinite res*) or divine: God. For Descartes, the thinking self is not perfect, but possesses the idea of perfection. His line of reasoning was as follows: knowledge of the idea of perfection, beyond my own imperfection, cannot come from myself, because I am imperfect and what I see is imperfect; rather it must come from a being that is more perfect than I, the creator of this innate idea. The *res infinite* is an uncreated substance, that thinks and that is the cause of all created beings. God is an eternal, immutable, independent, omniscient, omnipotent substance. The main attribute of this substance is obviously infinity.

The third is the extended substance (*res extensa*), represented by material things. This substance is a fundamental attribute of extension, and contains a triple dimension: shape, position and movement. According to Descartes, the soul is defined by thought and the body is defined by extension, so they are two separate things. Consequently, it is the soul that perceives and suffers the passions (desires, sadness, anger etc.) and the body is reduced, thus, to a machine governed by the laws of physics.

According to Descartes, the idea of infinity has been imposed by a nature that is higher than human, and can only come from this nature being infinite, so he interprets that the existence of infinity confirms the existence of God.

ZENO OF ELEA AND THE INFINITE PARADOX

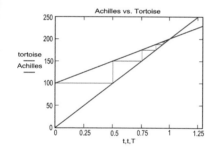

Zeno of Elea was a Greek philosopher belonging to the Eleatic school. His exact dates of birth and death are not known, but are usually said to be between 490 and 430 BC. He was a disciple of Parmenides, and is remembered for the broad conceptual arsenal with which he defended his master's thesis.

Zeno of Elea is known for his paradoxes or aporias, especially those that deny the existence of motion or the plurality of being. Zeno tried to prove that being is homogeneous, unique, and therefore, that space is not composed of discontinuous elements, but the entire cosmos or universe is a single unit. His method was the indirect proof of a thesis by *reductio ad absurdum* of the opposite view.

The arguments of Zeno are the oldest evidence that remains of infinitesimal thinking, developed many centuries later in the application of infinitesimal calculus that would be advanced by Leibniz and Newton in 1666.

Zeno contended that the sensations we get from the world are illusory, and specifically that there is no movement. One of his most famous paradoxes is the runner in the stadium, according to which a person cannot run the length of the stadium, as the runner must first reach the middle of it, but first has to reach halfway to the middle, and so on to infinity. Thus, theoretically, a person cannot run that distance, though the senses show that it is possible.

What is true, as demonstrated by the Scottish mathematician James Gregory (1638–1675), is that a sum of infinite terms can have a finite result. Obviously this modern interpretation of calculus was unknown in the time of Zeno.

In the paradox of Achilles and the tortoise, Achilles understands that after giving the tortoise an advantage he can never catch him, because to do so he must cover half of the distance that separates him from the animal, and beforehand, half of the half, thus always being left behind.

EPICURUS, THE VOID AND THE INFINITE

Epicurus of Samos (341 BC–270 BC) was a Greek philosopher, founder of the Garden, where all social classes of people were admitted, including women and slaves, which was quite shocking for the time. Of his numerous writings, only three letters have been preserved and some fragments, collected by Diogenes Laertes. The main sources of the doctrines of Epicurus are the works of Roman authors Cicero, Seneca, Plutarch and Lucretius, whose poem *De Rerum Natura* describes Epicureanism in detail.

The philosophy of Epicurus can be divided into three parts, the Canonic, which addresses the criteria by which we distinguish the true from the false, the Physics, the study of nature, and the Ethics, which is subordinate to the first two. Broadly speaking, he states that there is no more than one reality, the sensible world, thus he denies the immortality of the soul and suggests that like everything else, it is made of atoms. He defends rational hedonism in ethical theory and rejects the significance of politics, preferring a simple and self-sufficient lifestyle focused on the pursuit of happiness, in which friendship is the key. The pleasures of the mind are superior to those of the body, and both must be met with intelligence, trying to reach a state of spiritual well-being he called ataraxia. He argued that in nature there is no need for intervention by the gods and that natural phenomena could be explained by natural causes, more credible and acceptable than myths.

Epicurean physics uses the atomism of Democritus, but with modifications. The two basic principles are "out of nothing comes nothing" and "the All must be infinite both in respect of the number of atoms and in respect of the extent of void." All reality would consist of two basic elements, atoms and the void, which is the space in which these atoms move. The bodies are "systems of atoms." The number of atoms is infinite, as is the empty space, thus admitting the possibility of the existence of an infinite number of worlds as well as ours, which are born and perish, but the whole universe is eternal and imperishable. He suggested that the atoms are free and move with total spontaneity, an idea similar to the uncertainty principle of quantum mechanics.

Epicurean physics means that nothing comes into existence from nothing, while nothing destroys and is converted into nothing, meaning that everything is immutable and eternal.

SPINOZA AND THE THEORY OF THE INFINITE MODES

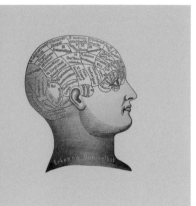

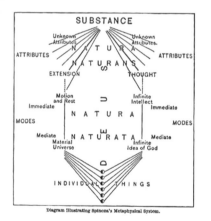

Diagram Illustrating Spinoza's Metaphysical System.

Baruch Spinoza (1632–1677) was a Dutch philosopher of Sephardic origin, and considered one of the three great rationalists of the 17th-century philosophy, along with René Descartes and Gottfried Leibniz. His major works include the *Tractatus de intellectus emendatione* and the fundamental *Ethica more geometrico demonstrata*.

In Part I of the *Ethica*, he introduces some concepts like "substance," "attributes" and "modes." The first notion refers to reality, cause of itself and also of all things, whose essence involves existence. Spinoza calls this reality God and nature, or more accurately, "God or nature," according to his famous phrase *Deus sive natura* .

This reality of which Spinoza speaks is an infinite being with infinite attributes, each of which is also of infinite concretions or modes. Thus defined, reality is eminently positive: it is essential, infinite, eternal by necessity. Reality is complete, it occupies everything that is and cannot have limitations, so the attributes of the substance are infinite. By attribute, he understands that the mind perceives a substance as a constitute of its essence. The more reality or being that one thing has, the more attributes it has.

God (or nature) and the world, its production, are then identical. All physical objects are the ways of God contained in the attribute "extension." Similarly, all ideas are the "modes" of God contained in the attribute "thought." The things or modes are nature made manifest, while the substance, or God, is nature brought into being, whose existence is necessary and eternal.

Spinoza distributed the modes into two systems, according to the attributes of "extension" and "thought," and classified them into infinite and finite. Infinite modes are subdivided into immediate and indirect. Thus, as for the attribute extension, the immediate infinite mode is movement and repose, the indirect infinite mode is the face of the whole universe, and the finite modes are the bodies. For the attribute thought, the immediate infinite mode is the absolutely infinite understanding and finite modes are particular ideas (including both true and false thoughts and notions of all kinds).

Spinoza's famous formula Deus, vive Substancia, sive Natura *expresses a notion of God as an equivalent of nature. Nature is a whole, a single substance, and things are merely parts of that whole infinite.*

mse?

ARISTOTELIAN PHYSICS OF INFINITY

Aristotle of Stagira (384 BC–322 BC) is one of the leading philosophers of the West. Son of Nicomachus, the physician of the Macedonian king Amyntas II, he entered the Academy of Plato at 18 years of age, where he remained about 20 years, until his master's death. He decided to leave Athens and go to Mitelene, where he was commissioned by Philip of Macedon to educate his son, Alexander. After Alexander the Great ascended the throne, Aristotle returned to Athens and founded a school near the Lycean Apollo temple, from which it took the name Lyceum. Upon the death of Alexander (323 BC), Aristotle was exiled to Chalcis where he died.

Among the works of his monumental legacy we will focus on *Physics*, a treatise in eight volumes, for it is here that we find the greatest number of references to infinity. Like other works of Aristotle, it is the result of the reconstruction work carried out by Andronicus of Rhodes probably around the first century BC, from pieces written by Aristotle at different times about physics. Aristotle is the first of the Greek philosophers who dealt with the concept of infinity and its possible existence. Previous philosophers had mentioned infinity in discussions of other topics, without providing an actual definition of infinity.

Aristotle addressed the issue from two angles: the infinite as a process of growth or final subdivision without end (infinity as a power) and the infinite as a whole or unity, truly unlimited (infinite as an act). The notion of potential infinity is focused on the possibility of proceeding always further without there being a last element in the endless recursion. However, Aristotle refused to accept the existence of infinity as an act. He rejected the physical existence of the infinite while he recognized the mathematical existence. He claimed that the world we know in action has limits, and the infinites we know as the infinite divisibility of space and the infinite series of numbers, are only potentials. An iterative process without end, such as the generation of natural numbers by adding a unit to the last number, is an infinite potential only because at any moment at which we stop, there is only a finite number of objects. There is no number by which adding to it another equals infinity, there is no actual infinite number.

According to Aristotle, we cannot conceive natural numbers as a whole. However, they are potentially infinite because we can always find a higher finite set. His distinction between actual infinity and potential infinity has fueled a debate that has extended throughout the history of philosophy and mathematics.

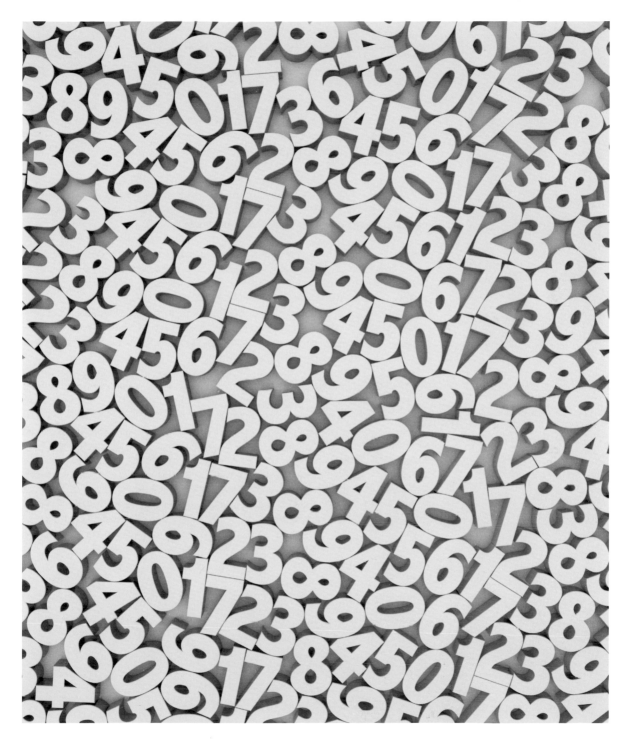

VOTAIRE AND INFINITY

François-Marie Arouet (1694–1778), known as Voltaire, was a French writer and thinker who was one of the leading representatives of the Enlightenment, a period that emphasized the power of reason, science and humanism.

In his reflections, Voltaire suggests two kinds of infinity, infinite duration and infinite space. He questioned whether, as a finite being, man can assume an exact idea of what infinity is, describing this conflict as "What is that which is eternally going on without advancing—always reckoning without a sum total—dividing eternally without arriving at an indivisible particle?"

According to the French thinker, it is impossible for infinity not to exist. It has been proven that infinite duration has elapsed. To start over is absurd, because nothing can start something. This demonstrates an infinite duration. In his thoughts, Voltaire provides the following reflection: "I distinguish between two eternities, the one before, the other behind me. When however I reflect upon my words, I perceive that I have absurdly pronounced the words. For at the moment that I thus talk, eternity endures and the tide of time flows. Duration is not separable and as something has ever been, something must ever be. The infinite in duration then is linked to an uninterrupted chain. This infinite perpetuates itself, even at the instant that I say it is passed. Time begins and ends with me, but duration is infinite."

Voltaire also suggests an infinity of space, but not before questioning the concept of space itself. Is it a being or is it nothing? No one dares to call it "nothing," nor do they know how to define it, but man knows that space exists. Our intelligence fails to understand either the nature of space or its end; we call it "immense" because we do not know how to measure it.

Voltaire argued that the difficulty of understanding the infinite time is due to the finite character of human nature.

INTUITIONISM

In the late 19th century, the various paradoxes and discoveries about infinity aroused great interest in philosophical foundations and mathematical logic. Intuitionism arose as a mathematical trend to confront and reject part of some truths inherent in classical reasoning, so it fits into the so-called "alternative logics."

This mode of reasoning was utilized by the Dutch mathematician L.E.J. Brouwer (1881–1966), who argued that logic does not precede mathematics, but depends on it. Within Brouwer's mathematical conception, objects and mathematical axioms are as they are intended, that is, born of the intuition of human thought. This dogma causes intuitionism to reject the majority of classical mathematics, as it denies everything that is not mentally constructible. The clearest example is the infinite.

Intuitionism does not support the idea of infinity in terms of the simultaneous presence of a multitude of objects or spaces (actual infinity), but is capable of supporting the infinite potential of the universe, that is, the possibility of generating objects indefinitely. Thus, intuitionism attempts to solve the problems that the existence of the infinite rejecting its existence poses for the arithmetic calculation. The fundamental thesis of this mathematical intuition is based on the assertion that mathematics is exclusively made up of a set of entities built intuitively by the same mathematician on which others will continue to build a clear, precise and fruitful operational system.

Mathematical objects are recognized as those that can be built, the intuitionist position refuses to accept the existence of the infinite but admits the possibility of infinite in potential, that is, given a set, you can build another with more elements.

$$k_3 = hf(x_{i-1} + \frac{h}{2}, y_{i-1} + \frac{k_2^{(i-1)}}{2})$$

$$\frac{b_i - (\sum_{j=1}^{i-1} a_{ij} x_j^{(k)} + \sum_{j=i+1}^{n} a_{ij} x_j^{(k)})}{a_{ii}}$$

$$\Delta y_i = \int_{x_i}^{x_{i+1}} y' dx \qquad \frac{b_i - (\sum_{j=1}^{i-1} a_{ij} x_j^{(k)} + \sum_{j=i+1}^{n} a_{ij}}{a_{ii}}$$

$$\int_{x_k}^{x_{k+1}} f(x, y) dx = \int_{x_k}^{x_{k+1}} y' dx = y(x)$$

$$k_2 = \sqrt{(y_n + 0.5\tau k_1)^2 + (t_n + 0.5\tau)}$$

THE TAO

道

Among the multiple forms of Chinese philosophy lies Taoism, whose origin dates back to its founder, Lao Tzu (4th or 5th century BC), author of *Tao Te Ching*, a seminal work of the Chinese culture that has managed to penetrate the political, religious and philosophical spheres of the East.

Tao Te Ching is a treatise organized into two books of 37 and 44 chapters in which Lao Tzu expresses his wisdom in a simple and concise way, highlighting the naturalness and spontaneity of man and reflecting the mystical aspect of Chinese tradition, the environmentalism of his followers and free movement of the spirit.

Taoism establishes the existence of three forces: one passive, yin, one active, yang, and a conciliatory, the Tao. The first two simultaneously oppose and complement each other; they are absolutely interdependent and function as a unit. The Tao is a greater creative, infinite, indeterminate and chaotic force, consisting of the unity of all beings and ways. For Taoism there is only one truth and it is that the Tao is infinite, therefore nature is infinite, a fact that makes the universe perpetuate in an infinite time that occurs in a loop of creations and destructions.

La Tzu described Tao as a natural law of the cosmos through which every person can exceed the limit of life and thus attain immortality, but it should be noted that here he does not refer to a Western-style immortal soul, but to control the body's tendency to decay. Taoists interpret "immortality" as the self-improvement of every being in community with the environment, and it was said that people living in harmony with nature were immortal. For this reason, some people interpret Taoism as the worship of nature as a supreme power, turning its followers into the first environmentalists.

Taoism greatly influenced other beliefs, especially Chinese Buddhism and particularly in terms of Chan meditation, better known in the West by its Japanese school, Zen Buddhism.

Use

COPY

SCHOPENHAUER AND WILL

The philosopher Arthur Schopenhauer (1788–1860) abandoned German idealism, had a significant influence on the art of the Romantics and was one of the pillars on which pessimism was based; the philosopher Friedrich Nietzsche found inspiration in his work as well. His major work, *Die Welt als Wille und Vorstellung* (*The World as Will and Representation*), provides the true synthesis of all his thought, characterized by the diversity of his influences: Buddhist and Hindu Eastern philosophies, Plato and Kant's doctrine.

For Schopenhauer, the reality of things is represented by a metaphysical principle called will (*Wille*). Will is an irrational force that integrates all of nature and the universe with all entities and beings therein. All the energies of nature express will, including natural forces and motivations and instincts.

Schopenhauer's pessimism is expressed in his conception of an infinite, unique and indivisible will, inspired by Romanticism. Man, in terms of individuals aware of that will, is bound to suffer. According to Schopenhauer, life is an eternal range between desire and boredom.

Will is infinite effort, an unlimited drive, therefore it can never bring satisfaction or a state of tranquility. What we call happiness is only a temporary cessation of desire. Desire, as an expression of the need and sense of deprivation, is a form of pain. Happiness is the freedom from pain, overcoming the need, but before long it turns into boredom: desires are reborn and thus this cycle perpetuates *ad infinitum*.

In 1809, Schopenhauer started at the University of Göttingen as a medical student and he went on to complete several courses but his passion for Plato and Kant directed his interests toward philosophy, classic philology and history.

SKIP negative view of life

THE ETERNAL RETURN

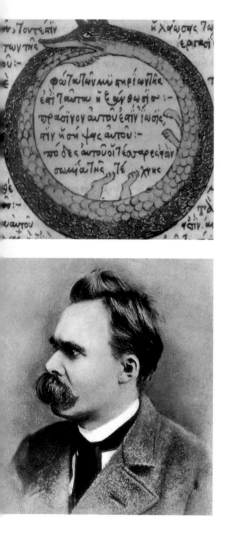

The eternal return is a philosophical doctrine that explains the history of the world and personal development as a process that is repeated. In eternal return, as in a linear view of time, events follow rules of causality. There is a beginning and an end of time that in turn rebuilds a principle. However, unlike the cyclic view of time, there are no cycles or new combinations of other possibilities, but rather the same acts repeatedly occur in the same order as they occurred, without any possibility of variation.

This concept has a precedent in Hindu thought and was adopted by the Greeks (especially Heraclitus and the Pythagoreans). It was subsequently formulated by the German philosopher Friedrich Nietzsche (1844–1900), who in his work *The Gay Science* argues that not only events are repeated, but also thoughts, feelings and ideas. This idea was taken up later in *Thus Spoke Zarathustra*.

In his book *The Gay Science* he proposes the eternal return as follows:

> "What if a demon were to say to you: This life as you now live it and have lived it, you will have to live once more and innumerable times more. And there will be nothing new in it, but every pain and every joy and every thought and sigh and everything unutterably small or great in your life will have to return to you, all in the same succession and sequence. (...) Or how well disposed would you have to become to yourself and to life to crave nothing more fervently than this ultimate eternal confirmation and seal!"

According to Nietzsche, this possibility terrifies us because we live life without the necessary intensity. However, there is another interpretation. Nietzsche presents eternal return as a representation of a view of life: life should be so intense and so perfect that we do not want anything to change. This would be an ethical doctrine in which there is no place for repentance: what you want, love it so you also want its eternal return.

Friedrich Nietzsche, a follower of Schopenhauer's way of thinking and Wagner's music, stood out for his radical critique of religion.

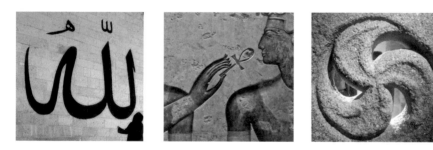

SYMBOLOGY

THE TIBETAN BUDDHIST ENDLESS KNOT

The endless or eternal knot is a symbol used in Tibetan Buddhism. Also called the "mystic dragon," it represents the infinite wisdom and compassion of the Buddha for all living beings.

It consists of a symmetrical, balanced and elegant knot, without a beginning or end, a closed knot that signifies eternity and unity. The same motif can be found in many representations of Chinese art as one of the eight auspicious symbols, illustrating infinity and longevity. It bears some resemblance to the traditional Celtic knot, and the same design scheme consisting of intricate closed geometric lines can be found in many cultures around the world, which makes it a near a universal symbol.

For Buddhists, this knot is a reminder that the universe is connected, and that many future events are rooted in the present. It has received many interpretations related to the interdependence, the pairs of concepts that depend on each other, such as wisdom and compassion, religion and the secular, tranquility and action.

In Tibetan tradition, the endless knot is a symbol of the ceaseless change of events, the network of lines reminiscent of the way in which the phenomena are interconnected in a closed cycle of cause and effect, in other words, karma. It represents the union of wisdom and the method. In tantra, it symbolizes the union of the female energy with the male and its harmonious union expresses the infinite love, infinite life.

The Tibetan knot is one of the eight auspicious symbols. With no beginning or end, it symbolizes the infinite wisdom of Buddha.

MANDALAS

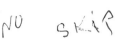

The word "mandala" is of Sanskrit origin and generally means "holy or magical circle." The mandala is a concentric and symmetrical figure in which everything is equidistant and is subordinated with respect to an axial point. They are used for their mystical and symbolic meaning in both building and in Buddhist visual productions. Thus, temples were built according to a mandala-based floor plan, generally square, divided into eight sections, each devoted to a divinity, such as the temple of Borobudur in Java.

The mandala, of which you can find parallels in medieval Europe and in other cultures and civilizations, has had a strong influence on the contemporary Western world because of pro-Eastern religious and philosophical trends in recent years. It is likely that the universality of these figures is due to the fact that concentric shapes suggest the idea of perfection and that the perimeter of the circle evokes the eternal return of the cycles of nature, as suggested in classical times by Hellenistic tradition.

This "circle" is a representation of the union of the individual with the cosmos, with the infinity that surrounds it, and of integrity, totality. The mandala originates from the central point and is built outwards, and could continue and expand to infinity. The fact that it is closed makes it represent a safe place, because the line around it symbolizes a protective barrier.

Over time, mandalas have been a tool for meditation and prayer, and a tool for diagnosis and monitoring of certain psychological disorders. The psychiatrist Carl Jung (1875–1961), who is largely credited with Western interest in mandalas, found that their properties were beneficial in psychotherapy: drawing mandalas, his patients began to bring order to their inner chaos. Today they are used to relieve stress and anxiety, and educational experts agree that working with mandalas helps to develop intuition, creativity and express thoughts and emotions that perhaps could not be expressed otherwise. Usually they are colored first from the center, which represents the individual, outwards which represents infinity, but it is also possible to do so the other way around, from the everything, the abstract to the individual and specific, yourself.

Carl Jung brought mandalas to the West insisting that by using them we can express our inner selves and analyze our personal situation.

THE LABYRINTH

A labyrinth (from the Latin *labyrinthus*, and Greek λαβύρινθος *labýrinzos*) is a route made up of streets and crossroads, with an ingenious and complex structure, whose design variations are endless, especially in the case of the rhizome labyrinth, which contains infinite possible variations. This type of labyrinth has no center, no periphery, no exit and the trails are interconnected.

Labyrinths are classified into two groups according to their relationship with the center and exit. First of all, there is the classic or unicursal labyrinth: It is one path covering the entire space to reach the center, that is, it does not offer the possibility of taking alternative routes, no forks, but there is one exit, which is also the entrance. As there is only one path to follow, you cannot get lost. The second group of labyrinths are mazes, labyrinths with alternative paths, there is both a right and wrong way to proceed to the exit.

Square or rectangular labyrinths are the oldest, the first known representation of a labyrinth of this type is found on a tablet from Pylos and tombs from ancient Egypt. Circular labyrinths appeared in the late 7th century BC in Etruscan Italy and later, they could be seen on coins of Knossos, in the late 3rd century, which would represent a map of the mythical Cretan labyrinth built by Daedalus to hide the Minotaur.

At present it is thought that the myth of the labyrinth is based in the palace of Knossos. A construction as sophisticated as this palace, replete with multiple rooms and all technological developments known at the time (for example, a sewer system) must have appeared intricate to the Achaeans. This thesis is supported by the fact that drawings have been found in the palace of Knossos of double-edged axes, *labrys* in Greek, which would have lent the name to the building.

Built in Crete about 1700 BC, the architectural complex composing the palace of Knossos, with its many parts and interlocking corridors, formed the maze that the legend describes as the Lair of Minotaur.

AZTEC SUN STONE

A key component of Aztec mythology and one of the most representative icons of the Mexican people, the Sun Stone was discovered in the late 18th century in Mexico, during the construction of the new cathedral in what is now the plaza del Zócalo. The exact original location of this monument has not been determined, but we know that it was in the main square of Tenochtitlan, where the Great Temple stood along with the main places of worship and political power. It is a circular basalt of 11.77 feet (3.6 m) in diameter and weighing over 48,000 pounds (24 tons), carved and crafted with unrivalled art.

The Sun Stone, often called "Aztec Calendar," might not be a calendar but a memorial stone of a sacred date, a ritual feast held every 52 years: the feast of the New Fire. In particular, the Sun Stone was engraved with the date that marks the celebration of the New Fire in 1479.

This monumental stone features elements related to the passage of time. Its design consists of a central image surrounded by five concentric circles. In the center, there is the face of Tonatiuh (Sun God), decorated with jade and holding a knife in his mouth. The four suns or previous eras are represented by square-shaped figures flanking the fifth sun, the current era, in the center. Below is the circle of 20 days, which corresponds to the representation of a month (the calendar had 18 months, 20 days each, plus 5 *nemontemi* or unlucky days). Next to it is the circle with the four cardinal points and the sun.

In the outer circle, two fire snakes with their mouths open are joined together. It is thought that they are Tonatiuh, the sun god, and Xiuhtecuhtli, the god of fire, that here symbolize the starry night sky and the night-earth place, where the sun sets, or perhaps a representation of the Milky Way, the galaxy containing our solar system. For the Aztecs, the Milky Way represents the largest expansion force with regards to man, before arriving at absolute totality.

The Aztecs dominated central and southern Mexico from the 14th to the 16th century, and established a vast empire that stood out for its excellent organization. Around 1325 a great city, Tenochtitlan, was founded, which became the capital (now Mexico City).

SKIP

OUROBOROS

Ouroboros, from the Greek *ourá*, "tail," and *borá*, "eating," is an ancient symbol depicting a serpent creature that swallows its own tail, forming a circular body. In some ancient representations it appears with the Greek inscription εν το παν, "everything is one." It has also been depicted by two snakes biting each other.

It represents the eternal return and other concepts perceived as cycles that begin again as soon as they end, destruction which in turn becomes creation. In a broader sense it symbolizes time and continuity of life. The image of the circle that forms the body of the animal is a clear metaphor for the cyclical repetition, for eternity.

Ouroboros is a symbol that different beliefs and civilizations have used to show an eternal universe, where everything always changes to return to its origin. The earliest representations of this symbol date back to Ancient Egypt and Ancient Greece. Some Ouroboros date back to the hieroglyphics found in the sarcophagus chamber of the pyramid of Unas, in 2300 BC. It also can be found in Norse mythology in the form of the serpent Jormungand, who grew so large that it could encircle the world and catch its tail with its teeth. In the practice of alchemy it expressed the unity of all things, material and spiritual, which never disappear but change form in an eternal cycle of destruction and creation anew, and equally it represented infinity. It can also mean the two dimensions of the world (represented by the interior and exterior of the circle), an interpretation devised by Christianity, where in addition the snake as an image of temptation suited perfectly: outside of the serpent, in the sky, there is no sin, but everything inside belongs to the kingdom.

The traditional representation of ouroboros in the different civilizations consists of a snake or dragon biting its tail and creating an endless circle.

THE BORROMEAN RINGS

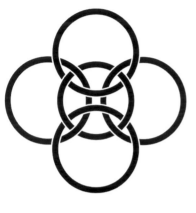

The Borromean rings consists of three linked rings with a very special property, which has a symbolic meaning: by cutting any of these three rings the other two are released, however the three together cannot be separated. This ring is actually impossible to build using flat rings, but it can be created when we use triangles or squares.

This name comes from the illustrious family Borromeo. The coat of arms of the dynasty in Milan, in fact, consisted of three rings in a clover shape symbolizing the triple alliance with the Visconti and Sforza families.

Yet it first appeared in the heraldic shields of the 15th century, long before Borromean rings existed, such as in Afghan Buddhist art in the 2nd century of the Christian era, or in the Valknut symbol of the Scandinavian myths in the 7th century. We can also observe some representations in Greek mythology.

Borromean rings have been used in different contexts to symbolize strength and unity, especially in the arts and religion. One example is the use given in Christianity during the Middle Ages; it was adopted to explain and illustrate the mystery of the Holy Trinity, in which Father, Son and Holy Spirit are one God. Similarly, the Celtic trisquel has also been represented as a symbol of the unity of the Knights of the Round Table in the Arthurian legends.

The rings represent the noble families of Milan, the Visconti, the Sforza and Borromeo, who formed an inseparable bond sealed by marriage.

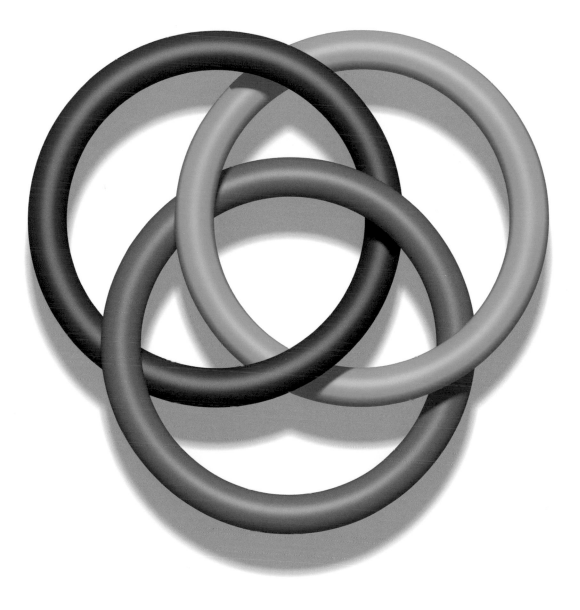

THE EGYPTIAN HIEROGLYPHIC EYE OF HORUS

Around 2700 BC, ancient Egyptians introduced the first system of unit fractions, fractions of numerator 1 and whose denominators are positive integers, for agricultural measurements of surface and volume. These fractions were represented by a hieroglyphic spelling of the Eye of Horus.

According to legend, Horus, the posthumous son of Osiris, the god who was murdered by his brother Seth, sought revenge for the death of his father and challenged his uncle. A terrible battle broke out in which Seth tore out Horus's eye; he cut it into 6 pieces and scattered them throughout Egypt. The assembly of the gods instructed Thoth, the supreme teacher of arithmetic, to gather together the parts and rebuild the eye.

The parts gathered by Thoth represented the fractions $\frac{1}{2}$, $\frac{1}{4}$, $\frac{1}{8}$, $\frac{1}{16}$, $\frac{1}{32}$ and $\frac{1}{64}$, whose sum does not reconstruct the unit, but make $\frac{63}{64}$.

The fraction that was missing to make 1 has, over time, acquired a different value throughout history. Some people believe that Horus lost it in his fight against Seth, and others understand that the missing $\frac{1}{64}$ represents the magic used by Thoth to restore the eye, proving to be an act of faith.

However, the most interesting theory relates the missing $\frac{1}{64}$ with the quest for infinity. Why not continue the search for the missing portion from the sum of the halves of the previous section? Indeed, if the series that Thoth suggested continues by adding half of the half of the half in continuous fractions, the result will undoubtedly be infinite, and this may be the real result of what the Horus missing pupil represents.

The ancient Egyptians calculated using unit fractions (numerator 1) such as $\frac{1}{2}$, $\frac{1}{3}$, $\frac{1}{4}$, ...
To express a fraction with a numerator other than 1, the Egyptians wrote it as a sum of distinct unit fractions, hence the sums of unit fractions are known as Egyptian fractions.

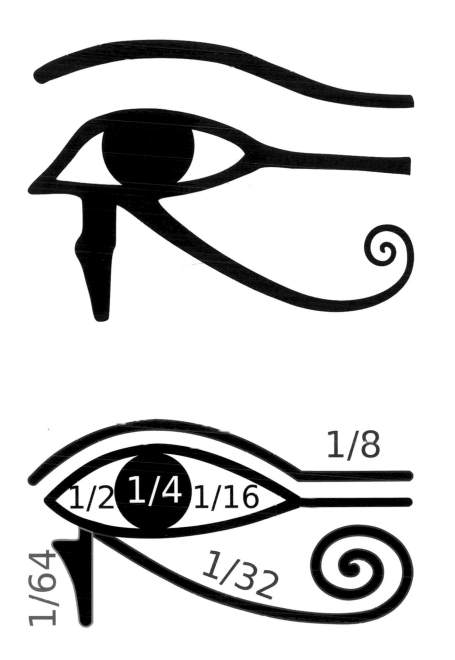

Use??

photo

COPY

THE CELTIC TRISQUEL

One element common to all Celtic people was the Trisquel, a sacred symbol reserved for the Druids, who functioned as priests of the Celtic spiritual beliefs, although their role covered many aspects, such as administration of justice and the study of the arts and knowledge.

The word "Trisquel" comes from Breton Celtic and means "three wings." The symbol consists of an outer circle representing the world and the infinite, within which there are three spirals with double turns that form three circles. These spirals are born from the same point, each symbolizing one of the component parts of the universe. The number three had a special significance for the Celts; it represented perfection and balance.

Its meaning is not entirely clear, but it generally relates to solar or astral worship, the beginning and the end, the eternal evolution and perpetual learning, and it could be considered a symbol of movement and change. It is believed that it could also symbolize the three primal forces of creation or the three streams of life that emerge from the primordial ocean.

Over the years more meanings have been attributed to the Trisquel. Since the Trisquel is also related to the divine triads of Celtic mythology—the balance between the heart, the body and the spirit, because the three spirals join the external circle which symbolizes the being and its relation with all the elements—with the advent of Christianity in the Celtic lands, the Trisquel became a representation of the Christian Trinity.

Other cultures such as Chinese, Hindu and Hebrew use symbols similar to the Trisquel.

SYMBOLOGY

ISLAM AND THE INFINITE

To understand the full extent of what Allah means to Muslims, it is essential that we have as correct as possible an idea of what the notion of infinity in Islam involves.

The presentations of the fundamentals of Islam always begin with a section devoted to Tançîh, the purity of Allah. The Tançîh emphasizes the indefinable character of Allah, his unintelligibility, his status as a challenge to understanding. To begin with, the reader is faced with the infinite, which is the response that man has for the secret of his own restlessness and uneasiness. The vastness of Allah, just as goodness and beauty of all that we love, is linked to the feeling of anxiety that the intuition of infinity produces.

One of the words related with the infinite is *al-Kull*, the All. The All is necessarily equal to the infinite: "all" implies that there is nothing outside, there is nothing outside of what is being mentioned. This evidence coincides with the notion of infinity, the unfinished, and not limited in any respect. Nothingness is foreign to it and what is excluded would be a limit, and what we call the infinite can no longer be, it would not be indelimitable. This idea follows the representation that Muslim makes of Allah, outside of which there is nothing, as he encompasses everything.

There is nothing, therefore, outside of the infinite, which is necessarily everything. It is therefore absurd to speak of a plurality of infinites or even an infinite thing (such as space, time or other concepts). Any particular thing, such as a determined thing, is not at the same time something else and, therefore, it is not everything: it is limited by its very nature and being what it is, it cannot be anything else. What we see in the world are well defined things, and therefore they are not Allah. Allah is a challenge to this perception.

Islam began with the preaching of Muhammad in 622 in Mecca (Saudi Arabia). Today there are over one billion Muslims in the world.

Use?

HEH, THE GOD OF ETERNITY

In Egyptian mythology, the Ogdoad is the set of four pairs of primordial deities that represented an inseparable entity. They acted together and embodied the essence of the primordial liquid chaos that existed before the creation of the world. The first pair is formed by Nun and Naunet, "the primordial waters" or "the chaos," the second pair is Heh and Hauhet, "infinite space" or "the unlimited," the third pair is Kuk and Kauket, "darkness," fourth, Nia and Niat, are "life," sometimes replaced by Tenemu and Tenemet, "the occult" or, later, by Ammon and Amonet, "the principle of mystery."

As we have seen, Heh, whose name means "infinity" (and its feminine form Hauhet), was the divine personification of the infinite, in the temporal aspect of eternity. Like the other elements of Ogdoad, the male form is depicted as a frog or a frog-headed human and the female form, like a snake or a human-headed snake. Its most common representation is embodied as a man kneeling on the symbol for gold, holding a palm stem in each hand (or only one) and sometimes with a stalk in his hair, as the palm represents long life for Egyptians. Representations of this form have a ring knotted at the base of each stem, symbolizing the infinite, and an ansate cross is often placed on his arm.

In hieroglyphics, the figure of Heh represented the figure of one million, which was considered almost equal to infinity in Egyptian mathematics. Therefore, this deity was also known as the "god of the millions of years," the symbolic measurement of eternity.

The pharaohs worshiped during their earthly life and hoped that, after death, the god of infinity granted them the possibility of eternal life.

LIMBO

In the Catholic tradition, limbo was a temporary place where the souls of good believers who died before the resurrection of Jesus (limbo of the patriarchs or *limbus patrum*) lived and the eternal place where those who die at a young age without having committed any personal sin, but without being absolved of original sin through baptism, were destined to spend eternity without torment nor glory (limbo of infants or *limbus puerorum*). The patriarchs escaped limbo when Jesus went down to hell, but the children would be doomed to remain there forever and not ever be able to enjoy the sight of God.

This idea became popular among Catholic believers in the Middle Ages, when it was first suggested that dead children who had not committed sins were to remain indefinitely at the edge of hell, but at a higher level where the fire would not reach them.

Nevertheless, the concept was never fully defined from a strictly theological point of view, and limbo has never been an official dogma of the Catholic Church. This idea arose as a solution to the problem of medieval theologians, who were in a dilemma with regard to the principle that those who died without receiving baptism that forgives original sin and all previous sins should go to hell. The idea that innocent children or the great patriarchs and other righteous men who lived before Christ were to share their fate with the damned for all eternity was unbearable and difficult to explain when preaching the idea of a merciful God. After Vatican II, the concept of limbo was abandoned and the current catechism trusts the fate of the unbaptized to the infinite mercy of God.

There are many literary works that have shaped limbo. These include the Divine Comedy, *the most famous work by Dante Alighieri, who said it was the first circle of hell and populated it with virtuous pagans and philosophers of the ancient world.*

KARMA: HE WHO SOWS, REAPS

Karma is interpreted broadly as a cosmic law of cause and effect. The Sanskrit word *karma* defines action, understood as something that triggers the entire cycle of cause and effect called *samsara*, as an immeasurable transcendent, invisible power, which derives from the acts of individuals. It is a fundamental belief in the doctrines of Buddhism, Hinduism, Yanism and other religions of Indian origin. Although these philosophies express differences in the meaning of the concept, they have a common basis for interpretation.

Karma explains human situations as a reaction to good or bad deeds carried out in a more or less immediate past. In Hinduism, the corresponding reaction is generated by the god lama, whereas in Buddhism and Yanism (where there is no controlling god), this reaction is conceived as a law of nature.

Under this doctrine, every individual has the freedom to choose between doing good or evil, but has to bear the consequences of their actions. Both for Hinduism and Buddhism, karma is not just about physical actions, but there are three factors that create reactions: acts, words and thoughts.

In accordance with the laws of karma, the behavior of an individual during their life will inevitably influence successive lifes, as a single human life is not enough to experience the full effects of the actions ("receiving" for all the good or "paying" for all the evil that has been caused). Buddhism affirms that there are no undeserved pleasures or unwarranted punishment, but everything answers to a universal justice. For Hindus, the individual essence of people adopts a material body several times throughout its existence. Depending on the merits or lack of them, the soul is reincarnated in a higher, medium or lower existence traveling from the most heavenly to hellish states, passing through human life. Samsara, escaping this ongoing process, is only possible after having amended the weight of the karma.

Under the law of karma, the effects of all actions, good and bad, create present and future experiences and make the person responsible for their own life.

SKIP?

THE PHOENIX

The phoenix is a mythical bird that after death is reborn, rising from its ashes. According to the myth, its size was similar to that of an eagle with feathers of a red, orange and yellow glow, and with beak and talons of extraordinary force. Archaic cultures that settled on the shores of the Mediterranean Sea tell different versions of the legend. The ancient Egyptians were the first to talk about the Bennu that the Greeks converted into the phoenix. Contrary to the myths of other civilizations, in the Egyptian myth it was neither predatory nor a tropical bird, but more like a crane. Also it did not emerge from fire but water.

A one of a kind and incomparable beauty, some variants tell that it rises from the ashes every millennium, others every five hundred years, but all agree that periodically it dies in flames. When the phoenix senses that it is approaching the end of its life, it collects twigs of sandalwood, frankincense, cardamom, cedar and other woods and herbs and makes a huge nest in the crown of a palm tree. In the end it launches, opening its splendid wings and letting the sun's rays ignite the nest, and it becomes consumed by the flames while singing a song of rare beauty and everything is reduced to fragrant ashes. Among the remains of the fire appears an egg that the sunlight incubates until, after 3 days, the shell breaks and the same phoenix emerges again.

Besides being a pagan symbol, it was also adopted by Christian mythology. In this myth, the phoenix was born in Eden, beneath the forbidden tree. When Adam and Eve were expelled from paradise, a spark fell from the flaming sword of the angel into the nest of the bird and set it afire. The phoenix was the only animal that had resisted the temptation and for its loyalty, it received the gift of immortality. Since then, it can be reborn from its own ashes.

This mythological creature is a symbol of spiritual and physical rebirth, purity and immortality. The Greeks gave it the name Phoenicoperus *(which means "red wings"), a term adopted by zoologists to designate flamingos.*

THE FOUNTAIN OF ETERNAL YOUTH

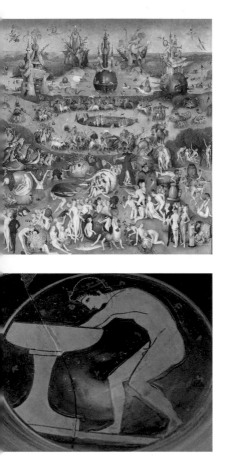

The Fountain of Youth is a legendary fountain, symbol of immortality and eternal youth, which appears in the classical and medieval mythology of many cultures. According to legend, the source, whose spring would be in the Garden of Eden, heals and rejuvenates those who drink its waters or bathes in them.

The location of the mythical source has been discussed since ancient times. The first legend narrated by Herodotus speaks of an unlocatable underground source located in Ethiopia. The ancient Greeks believed that the Ethiopians and in general, the inhabitants of Central Africa were very long-lived, and this story attempts to explain this fact.

Other stories related to a source of healing waters appear in the *Novels by Alexander*, that until the Renaissance many treasure hunters read to find precise directions. In the Oriental versions of these texts, the "water of life" is spoken of, a legendary source that can only be found after passing the "land of darkness," a mythical region of the Caucasus that was said to be home to monsters and spirits. The Arabic version of these romances by Alexander the Great were very popular in Spain during and after the Muslim era and were spread by the conquistadors who traveled to America.

Similar stories of miraculous waters widespread among indigenous peoples of the Caribbean fueled the myth during colonization. Native tribes spoke of the healing powers of the waters of the mythical land of Bimini. The legend became more popular in the 16th century, when it was linked to the Spanish explorer Juan Ponce de León. According to an apocryphal story that blends Eurasian and New World elements, Ponce de León discovered in 1513 what is now Florida, during one of the many explorations that were performed in pursuit of the fountain of youth.

For thousands of years, mankind has dreamed of rejuvenating spring water sources with elixirs capable of curing all diseases and bestowing immortality.

PHOTOGRAPHIC CREDITS

With the permission of Dreamstime.com: p. 8, © Solarseven; p. 9, © Philcold; p. 148, © Kts; p. 22, Universe spiral © Martin Bangemann; p. 27, Space Telescope © Xneo; p. 44, Human being asking Universe... © Cluc; p. 47, The future of science © Ioana Davies; p. 76, High speed laboratory centrifuge © Horia Vlad Bogdan; p. 77, Centrifuge © David888lee; p. 85, Peas isolated on White © Nailia Schwarz; p. 108, Inside a 3D Sierpinski sponge fractal object © G00b; p. 112, Vector background © Faustpr; p. 121, Möbius Strip © Martin Green; p. 125, © Etternity; p. 127, Nautilus Shell © Kuleczka; p. 145, Number 0 Background © Spaceheater; p. 177, Snow Flake © Pureguitarfury; p. 186, Silver DNA spiral © Ordogz; p. 188, © Oneo2; p. 191, Pen © Svit; p. 207, QR Code © Alesaggio; p. 214, Connect mind © Kts; p. 224, Stairs to papal apartments © Mirek Hejnicki; p. 237, DSLR Camera Spiral Twirl © Vlue; p. 259, Universe - Jupiter and a starfield © Cobalt88; p. 276, Manuscripts in Latin © Piccia Neri; p. 289, Endless knot symbol © Charon; p. 291, Mandala blue blackground © Youichi; p. 293, Labyrinth © Angelo Gilardelli.

With the permission of Flickr.com: p. 36, © woto; p. 50, Cesium atomic clock © mightyohm; p. 90, HeLa S3 © opiado; p. 92, At the beach © Boomer Depp; p. 100, Milnesium tardigradum - Dr. William R Miller © SaguaroNPS; p. 101, Minibiotus intermedius - Dr. William Miller © SaguaroNPS; p. 142, John Wallis © tonyntone; p. 155, © Claus Rebler; p. 160, L'Hôpital, illustrations from the book *Historia Universal* by César Cantú © Flickr/El Bibliomata; p. 173, © gadl; p. 184, © jurvetson; p. 184, Atmos © NeitherFanboy; p. 185, Atmos clock © Wm/Rama; p. 202, Rubik's Cube © Bramus!; p. 202, Rubik's Cube Collection © Scarygami; p. 203, Rubik's Cube © Cubmundo; p. 211, Weltzeituhr, Alexanderplatz, Berlin © kai hecker; p. 244, Roman Opalka © scalleja; p. 252, *Self Obliteration Part II* (1967) by Yayoi Kusama © Chicago Art Department; p. 252, *Infinity Room* by Yayoi Kusama © Cea.; p. 253, *Passing Winter* (2005) by Yayoi Kusama © Marshall Astor; p. 283, Statue of Lao Tzu in Quanzhou © tom@hk | æ¹¯ç±³tomhk; p. 284, The Colours of the Soul © h.koppdelaney; p. 300, Eye Of Horus © Nerdcoresteve; p. 303, Trisquel calado de Castromao © FreeCat; p. 305, sheikh lotfollah mosque © seier + seier; p. 314, The bennu bird © wm/Jeff Dahl; p. 314, *Bilderbuch für Kinder* (1790-1830) by Friedrich Justin Bertuch; p. 316, *The Garden of Earthly Delights* (c. 1490-1510) by El Bosco; p. 316, *Youth washing himself at a fountain* (c. 510-500 a. C.) © wm/Jastrow.

With the permission of Shutterstock.com: p. 11, © edobric; p.29, © Vadim Sadovski; p. 30, © optimarc; p. 31, © PSD photography; p. 34, © holbox; p. 35, © vilainecrevette; p. 37, © Ella Hanochi; p. 42, © Mike Heywood; p. 45, © red-feniks; p. 62, © pashabo; p. 64, © Tyler Olson; p. 65, © Vit Kovalcik; p. 67, © pixelparticle; p. 74, © EastVillage Images; p. 75, © Alexander Raths; p. 78, Berillium atom structure on the blue background © baldyrgan; p. 79, Illustration of an atom © MrJafari; p. 81, © Abstract background of a destruction brick wall © Rashevskyi Viacheslav; p. 91, HeLa (cervical cancer) cells in culture © Loren Rodgers; p. 93, Large Blue Surfing Wave Breaks in the Ocean © EpicStockMedia; p. 95, © DerLuminator; p. 97, Silhouette of a pipe with smoke © Sergey Panychev; p. 99, A pair of hands holding the Earth with the moon in the background © Shane Kennedy; p. 102, Optical prism refracting a ray of white light © Andrea Danti; p. 102, Color circle isolated with clipping path © Dario Sabljak; p. 103, Colorful tika powders on Indian market © Aleksandar Todorovic; p. 111, Abstract number concept © Pixel 4 Images; p. 114, © IrinaK; p. 115, Vintage photo, circa 1900 of a chimp typ © ChipPix; p. 122, Full Frame Neatly Coiled White Rope © diak; p. 123, 3D paper roll on a black background © Umberto Shtanzman; p. 128, Sunflower seed pattern © sarah2; p. 132, Very long corridor in a hotel © ID1974; p. 132, Door of a very important person © Wilm Ihlenfeld; p. 139, 3D rendered Molecule © Lorelyn Medina; p. 146, Two dice © Sergey Mironov; p. 147, Flipping a coin © Alexey Stiop; p. 148, Plastic toy bricks © Alexey Rozhanovsky; p. 150, Color building blocks as background © atm2003; p. 151, Code numbers © 1xpert; p. 171, Elegant abstract fractal background © Dragonfly22; p. 174, Steampunk engine and boiler © Paul Fleet; p. 175, 3D render of Escher's inspired stairs © Fotocrisis; p. 187, Original micro-photo of living dividing cells © Dimarion; p. 188, Monitor background © Taiga; p. 189, Tech Cubes © Michelangelus; p. 193, Light Bulbs © balein; p. 205, Nanobots © Andrea Danti; p. 215, World Wide Web concept © cybrain; p. 220, Breaking the barrier © Karl R. Martin; p. 223, Illusion of reality © XYZ; p. 227, A long endless music-sheet © Menna; p. 231, Fantasy Sleep Composition © Bruce Rolff; p. 236, Infinity image of LCD monitor © pryzmat; p. 242, Highway © Photosmart; p. 243, Railroad tracks parallel transport © apichart boonsin; p. 245, Random numbers © marekuliasz; p. 247, Infinite © nito; p. 258, © Iryna1; p. 310, © Magicinfoto; p. 258, Paris - Le Pantheon © Marco Cannizzarro; p. 260, Solar System © Morphart Creations Inc.; p. 261, Diagram of the planets in the Solar System © Jurgen Ziewe; p. 270, Running athlete at the stadium © Vladimir Wrangel; p. 277 Field of numbers © GrandeDuc; p. 279, Infinity time © Liseykina; p. 281, Mathematics background © Vasilius; p. 287, Man and the time © robodread; p. 309, Woman in dead tree © Elena Ray; p. 310, Fragment of Ancient Tibetan tangka © Zzvet; p. 310, Chiang Nai, Thailand © magicinfoto; p. 311, Tibetan Mandala © Nadina; p. 313, Phoenix through a red sky © Ellerslie; p. 315, Havasu river © Galyna Andrushko.

Other images: p. 10, Albert Einstein in Vienna, 1921 © Ferdinand Schmutzer; p. 10, © hardballtimes.com; p. 12, © lifewithheadinjury.wordpress.com; p. 14, © Alexis Monnerot-Dumaine; p. 15, © its authors; p. 15, The 4 paths to a number 4 © Wm/R. A. Nonenmacher; p. 16, © simplefire.wordpress.com; p. 16, © HowStuffWorks; p. 17, © Ævar Arnfjörð Bjarmason; p. 18, From the book *Oliver Heaviside: Sage in Solitude* by Paul J. Nahin; p. 18, © shade @ wiki.nl.; p. 18, © de.wikipedia.org; p. 19, © agri Erdogdu; p. 20, © jones.math.unibas.ch; p. 20, © its authors; p. 20, © jones.math.unibas.ch; p. 21, Cornu spiral © wm/Inductiveload; p. 23, M31: The Andromeda Galaxy © Robert Gendler; p. 24, © wm/Think Tank; p. 24, Copernican Universe, Thomas Digges; p. 24, Olbers' Paradox © 2011 Northwestern College; p. 25, Olbers' Paradox © wp/Wolfmankurd; p. 26, © NASA; p. 26, © wm/Brews Ohare; p.28, © from the article "Does the Universe Have a Mind?" by William Pepperell Montague in *The Saturday Review* (1947); p.28, © wp/"Professor Einstein's Visit to the United States", *The Scientific Monthly* (1921); p. 30, Detail of *Triumph of St Thomas* (c. 1365) by Andrea Bonaiuto © Zenodot Verlagsgesellschaft mbH; p. 32, K0-Kobar oscillation © wm/Bamse; p. 32, © wm/Harry Mustoe-Playfair; p. 33, © wp/TriTertButoxy/Sandbox; p. 34, Illustration of the ecosystem of *Posidonia oceanica* © its authors; p. 34, Fruit of *Posidonia oceanica* © wm/Tigerente; p. 36, © wm/Luca Antonelli; p. 38, From the book *The Pictorial Atlas of the Universe* by Kevin Krisciunas and Bill Yenne; p. 38, From the magazine *Astrophysical Journal* by J. H.Taylor & Cordes; p. 38, Rotation curve for Milky Way © wm/Brews Ohare; p. 39, A Roadmap to the Milky Way © NASA/JPL-Caltech/R. Hurt (SSC/Caltech); p. 40, © Archives of SIBA; p. 40, From *Popular Science Monthly*, vol. 33; p. 41, © www.spoki.lv; p. 42, © wm/Anynobody; p. 42, © wm/Xerxes314; p. 43, Illustration of Stellar Explosion of SN 2006gy © NASA/CXC/M. Weiss; p. 44, Hubble Ultra Deep Field Image Reveals Galaxies Galore © NASA, ESA, S. Beckwith (STScI) and the HUDF Team; p. 46, P.A.M. Dirac at the blackboard © its authors; p. 46, Dirac sea for a massive particle © wm/Incnis Mrsi; p. 48, *Principia Mathematica* by Isaac Newton; p. 49, Sir Isaac Newton, Sir Godfrey Kneller; p. 50, Illustration of space-time curvature © wp/Johnstone; p. 51, © Dahotsky; p. 52, Eukaryotes and some examples of their diversity © various authors, compiled by Vojtech.dostal; p. 52, © wm/Franciscosp2; p. 53, © Polyhedron; p. 54, © wp/XaosBits; p. 55, © wp/Wikimol; p. 56, Max Planck (c. 1930) © its authors; p. 57, Stars Spring up Out of the Darkness © NASA/JPL-Caltech; p. 58, © KIPAC/SLAC/M. Alvárez, T. Abel and J. Wise; p. 59, Kruskal diagram for an eternal black hole © wm/TimothyRias; p. 60, Bust of Edmond Halley at the Museum of the Royal Greenwich Observatory, London © wm/Kdkeller; p. 61, Comet P/Halley as taken March 8 © W. Liller, Easter Island, part of the International Halley Watch (IHW); p. 62, Otto Stern and Lise Meitner © wm/GFHund; p. 62, © sxc.hu/ Thomas Picard; p. 63, © sxc.hu/ Jeff Hire; p. 66, © NASA; p. 66, Heisenberg gamma ray microscope © wm/Radeksonic; p. 68, Chemical structure of DNA © Madeleine Price Ball; p. 69, An overview of the structure of DNA © wm/Michael Ströck; p. 70, A Lorentzian wormhole © wm/Allen McCloud; p. 71, Black Hole Grabs Starry Snack © NASA/JPL-Caltech; p. 72, © wp/Lunch; p. 73, © sxc.hu/Brad Harrison; p. 74, A molecular model of the bacterial cytoplasm © Adrian H. Elcock; p. 76, Tabletop centrifuge © wm/Magnus Manske; p. 80, © sxc.hu/John Nyberg; p. 80, © wm/DeepKling; p. 82, Hydrogen atom © Wm/Bensaccount; p. 83, Semi-classical electron orbits with principle quantum number n= 5 © wm/Meter Kuiper; p. 84, © sxc.hu/vancity197; p. 84, © wp/Tocharianne; p. 84, © wm/Magnus Manske; p. 86, Hubble finds dark matter ring in galaxy cluster © NASA, ESA, M.J. Jee and H. Ford; p. 86, © PD-USGOV-NASA; p. 87, Cosmos 3D dark matter map © NASA/ESA/Richard Massey; p. 88, From the book *Evidence as to Man's Place in Nature* by Thomas H. Huxley; p. 88, A Venerable Orang-outang from *The Hornet*; p. 89, Charles Robert Darwin © John Maler Collier; p. 90, HeLa cells stained with Hoechst 33258 stain © wp/TenOfAllTrades; p. 92, A particle motion in an ocean wave © wp/Vargklo; p. 92, Surface wave parameters © wm/Luis Fernández García; p. 94, Combustion of methane in dioxygen © Christophe Dang Ngoc Chan; p. 96, Carbon cycle-simple diagram © wm/FischX; p. 96, AIRS Map of Carbon Monoxide Draped on Globe © NASA/JPL; p. 100, *Gaea* (1875) by Anselm Feuerbach; p. 102, Light wave © wp/Heron; p. 106, Portrait of Gottfried Leibniz © Christoph Bernhard Francke; p. 107, © piratesandrevolutionaries.blogspot.com.es; p. 107, Secant line of a function © wikimatematica/Jpvillegas; p. 108, Hollow Menger Sponge fassade from behind © wm/Daniel Schwen; p. 109, Menger Sponge after four iterations © wm/Niabot; p. 109, Development row from the Menger Sponge © wm/Niabot; p. 110, © casio-schulrechner.de; p. 110, © es.wikipedia.org; p. 110, © de.wikipedia.org; p. 113, Full Color Theorem © wm/XalD; p. 113, Full Color Theorem © wm/Inductiveload; p. 114, Émile Borel, deputy for Aveyron © Agence Mundial; p. 114, © de.wikipedia.org; p. 116, Example of the set semantics of Term logic © wm/Dhanyavaada; p. 116, © de.wikipedia.org; p. 116, Sets - subset A of B © wm/Ed g2s; p. 117, Venn diagrams © Tilman Piesk; p. 118, Woodcut from *De Divina Proportione* by Luca Pacioli; p. 118, © artofraz.com; p. 119, Greek letter phi © wm/Dcoetzee; p. 119, From the book *Libri tres de occulta philosophia* by Heinrich Cornelius Agrippa; p. 120, August Ferdinand Möbius © Adolf Neumann; p. 121, Möbius Strip © wm/David Benbennick; p. 122, Archimedean spiral and parameters © wm/Kmhkmh; p. 124, *Self Portrait* by Albrecht Dürer; p. 124, Cut nautilus shell © wm/Onofrio Scaduto; p. 124, Approximate and true Golden Spirals © wp/Cyp; p.124, its authors; p. 126, © its authors; p. 126, Logarithmic spiral © wm/Kaboldy; p. 128, Hindu lattice © wp/Bilious; p. 128, Bust of Bertel Thorvaldsen at Thorvaldsen Museum © Stefano Bolognini; p. 129, A tiling with Fibonacci number sized squares © wp/Herbee; p. 130, Location map of the UK © wm/NordNordWest; p. 131, Partial view of the Mandelbrot set © wm/Wolfgang Beyer © wm/Wolfgang Beyer; p. 132, David Hilbert © its authors; p. 133, The first six steps of David Hilbert's space-filling curve © Zbigniew Fiedorowicz; p. 134, © its authors; p. 135, Povray rendering of Gabriel's Horn © wm/RokerHRO; p. 136, Rhumb-line spiraling towards the north pole © Alvesgaspar; p. 137, Distinction between loxodrome and orthodrome © wm/McSush; p. 138, Amedeo Avogadro © C. Sentier; p. 140, A Gosper Island, fifth iteration © wm/Inductiveload; p. 142, Bernoulli's Lemniscate © wp/rdrs; p. 142, Track left by a point with the constant variation of parameter t © wm/Framars90; p. 142, Bernoulli's Lemniscate © wm/Zorgit; p. 143, Bernoulli's Lemniscate © wm/Zorgit; p. 144, The hyperbola © Daleh; p. 150 © de.wikipedia.org; p. 152, Euclid of Megara © wm/ Justus van Gent; p. 152, Detail of a scene in the bowl of the letter P © its authors; p. 153, *Oxyrhynchus papyrus* (P.Oxy. I 29) showing fragment of Euclid's *Elements*; p. 153, X axis: n-th Mersenne prime number. Y axis: number of digits © wm/Sabbut; p. 156, Partial solution of the "Squaring the circle" task © Audriusa; p. 156, Dimensions required for squaring a unit circle © wm/Alexei Kouprianov; p. 157, © morgueFile/veggiegretz; p. 158, A portrait of Pierre de Fermat © its authors; p. 158 © agawebs.com; p. 159, © *Diophantus*, with notes by Pierre de Fermat; p. 160, © ca.wikipedia.org; p.160, © Getty/SCIENCE SOURCE; p. 161, © Bibliothèque Nationale de France; p. 162, Fermat spiral © wm/Sanbec; p.162, © Getty/Photo Inc; p. 163, Colored spiral painted on the sidewalk © sxc.hu/alexiares; p. 163, Pattern of florets in a sunflower following Vogles model © R. Morris; p. 164, © morgueFile/clarita; p. 164, Simple example of a Markov chain © wp/Fcady2007; p. 164, Graph of a Markov chain © wm/Joxemai4; p. 165, Game of Snakes & Ladders © Jain Miniature; p. 166, Uniform tiling of hyperbolic plane, x3x7o © Anton Sherwood; p. 166, Many parallel lines through the same point © wm/Jean Cristophe BENOIST; p. 167, sxc.hu/Lkay; p. 168, *The number of the Beast* (c. 1805) by William Blake; p. 168, Numbers 666 engraved on granite headstone © Karin Hildebrand Lau; p. 168, © upload.wikimedia.org; p. 169, *The Destruction of Leviathan* (1865) by Gustave Doré; p. 170, Newton-Fraktal der 3 © wm/LutzL; p. 172, King Otto IV of Brandenburg © Meister des Codex Manesse; p. 176, Koch snowflake © wm/Solkoll; p. 176, A Koch snowflake is the union of infinitely many regions whose boundary is triangle © Chas zzz brown; p. 177, Koch Curve © wm/Solkoll; p. 180, © web.mit.edu; p. 181, © wm/Minzinho; p. 182, © jouleunlimited.com; p. 183, © sxc.hu/MsDotty; p. 186, *The Fountain of Youth* (1546) by Lucas Cranach the Elder; p. 188, Infinity © Kakaopor; p. 190, International Space Station Imagery © nasaimages.org; p. 192, © centennialbulb/Dick Jones; p. 194, Silvered Dewar calorimeter © S. Szpak & P. A. Mosier-Boss; p. 195, Cold fusion electrolysis © wp/Pbroks13; p. 196, Illustration of a wind turbine © wm/norro; p. 197, From the book *Wind Energy in America: A History* by Robert W. Righter; p. 198, Overall view of the LHC © CERN; p. 198, Real CMS proton-proton collision events in which 4 high energy muons (red lines) are observed © CERN; p. 199, View of the CMS Detector before closure © CERN; p. 200, © YERT; p. 201, © YERT; p. 202, © freepixels.com; p. 204, TEM (a, b, and c) © wm/Nandiyanto; p. 204, 3D model of a en:C60 molecule © wm/Michael Ströck; p. 206, qrcode © wp/Brdall; p. 208, © Infinitec; p. 210, © Infinitec; p. 210, Celsium clock © wm/Zubro; p. 212, Pluripotent, embryonic stem cells originate as inner mass cells within a blastocyst © wp/Mike Jones; p. 213, From the book *Follow the Money. The Politics of Embryonic Stem Cell Research* by Nissim Benvenisty; p. 214, Isaac Asimov © Rowena Morrill; p. 216, Crystalline structure of graphene © wm/AlexanderAIUS © wm/AlexanderAIUS; p. 216, The Morph Concept © research.nokia.com; p. 216, Graphit-Gitter © Anton (rp) 2004; p. 217, © morgueFile/danielito; p. 218, F/A-18 Hornet Hornet breaks the sound barrier © Ensign John Cay (U.S. Navy); p. 218, A sound barrier chart © wm/Pbroks13; p.218, © Getty/Dorling Kindersley; p. 225, Stairs at the exit of the Vatican Museums © Eigenes Foto; p. 226, Richard Wagner © wp/Franz Seraph Hanfstaengl; p. 226, Leitmotif of Siegfried by Richard Wagner; p. 228, Self-portrait of Andrea Pozzo at Il Gesù Church, Rome © wm/Grentidez; p. 229, Jesuit Church Vienna ((Andrea Pozzo, 1703) © wm/Alberto Fernández Fernández; p. 230, *Eureka: A Prose Poem* by Edgar Allan Poe © Putnam; p. 232, Métronome gradué à battements muets Étienne Loulié (v.1637-1702) © wm/Hautbois; p. 233, Maëlzel Metronome © Myself; p.234, © wp/Google Art Project; p.234, © Leandros World Tour; p.235, © wp/WebMuseum at ibiblio; p. 236 © cakitches.com; p. 238, Count Giacomo Leopardi © A. Ferrazzi; p. 239, *Infinito: Secondo manoscritto autografo* © Vissio, Archivio comunale; p. 239, *The wanderer above the sea of fog* (1818) by Caspar David Friedrich © wm/Cybershot800i; p. 240 From the book *The Seaman's Secrets* by John Davis; p. 241, A modern nautical compass rose © wp/Mysid; p. 246, Jorge Luis Borges © its authors; p. 248, *Portrait of Niccolò Paganini* (1836) by John Whittle; p. 249, *Niccolò Paganini* (1831) by Richard James Lane; p. 250, Solomon's knot configuration © its authors; p. 251, Aquileia basiliek © wp/Pvt Pauline; p. 252, Installation at the Yayoi Kusama Special Exhibition © wm/Samuel Mark Thompson; p. 256, Anaximander, detail of *The School of Athens* (1510-1511) by Rafaello Sanzio; p. 256, Anaximander's models of the universe © Dirk L. Couprie; p. 257, Anaximander's lost map of the world © wm/Bibi Saint-Pol; p. 257, Anaximander's models of the universe © wm/Bibi Saint-Pol; p. 258, © its authors; p. 260, © its authors; p. 262, Immanuel Kant © its authors; p. 263, © Sven Geier; p.264, © Getty/DEA PICTURE LIBRARY; p. 265, © its authors; p. 266, Blaise Pascal (1623-1672) © wm/Janmad; p. 266, © sxc.hu/ugaldew; p. 267, Pascaline © wm/David Monniaux; p. 268, © its authors; p. 268, Pain pathway in *Traité de l'homme* (1664) by René Descartes; p. 269, © its authors; p. 270, © its authors; p. 270, How Achilles catches the tortoise © wm/Richard-59; p. 271, Reproduction of an engraving published in *Diogenis Laertii De Vitis* by Marcus Meibomius; p. 272, From the book *10.000 Meisterwerke der Malerei* by The Yorck Project; p. 272, *Reading Homer* (1885) by Lawrence Alma-Tadema; p. 273, Hermes-type bust of Epicurus © Eric Gaba; p. 274, Phrenologie © Friedrich Eduard Bilz; p. 274, © its authors; p. 275, Baruch de Spinoza © its authors; p. 278, Portrait of François-Marie Arouet, *Voltaire* © Catherine Lusurier; p. 278, *Diccionario Filosófico* by Voltaire; p.280 © E. van Moerkerken; p. 282, Lao Tse © Widodo; p. 282, From the book *The Secret of the Golden Flower*; p. 282, Chinese character for the word *tao* © its authors; p. 284, Arthur Schopenhauer © its authors; p. 285, From the book *10.000 Meisterwerke der Malerei* by The Yorck Project; p. 286, *Synosius* (1478) © wm/Carlos adanero; p. 286, F. Nietzsche © its authors; p. 288, Endless knot © wm/AnonMoos; p. 288, Tibetan endless knot © wp/Rickjpelleg; p. 290, Mandala © its authors; p. 290, © its authors; p. 291, Reverse of a clay tablet from Pylos © wm/Marsyas; p. 292, © Nordisk familjebok; p. 294, Scan of gold pendant replica of Aztec calendar stone © wp/Isis; p. 295, © Antonio de León y Gama; p. 296, Ouroboros symbol © wm/AnonMoos; p. 297, Ouroboros © wm/Nergal; p. 298, Borromean Cross © Christoph Sommer; p. 298, Borromean rings configuration © wm/AnonMoos; p. 299, The Borromean rings © wm/Jim.belk; p. 300, Horus presents the breath of life to the Pharaoh © sxc.hu/Nicole_N; p. 301, Eye of Horus © wm/Jeff Dahl; p. 301, Eye of Horus © wm/Benoît Stella; p. 302, Wheeled form of a Triskelion or Triple Spiral symbol © wm/AnonMoos; p. 302, A Guarda, Galicia © wm/Froaringus; p. 304, Muslim religion praying © omernraja; p. 304, Qur'an © sxc.hu/sumeja; p. 304, © wm/Nevit Filmen; p. 306, Heh and Hehet © wm/JMCC1; p. 307, An aspect of Heh © wm/Jeff Dahl; p. 308, *Christ in Limbo* by a follower of Hieronymus Bosch; p. 308, *Dante meets the four bards* (1857) and *Dante meets the unbaptised* (1857) by Gustave Doré.

wp = wikipedia | wm = wikimedia